β Beta

Multiple-Digit Addition and Subtraction

Student Workbook

Math·U·See®

1-888-854-MATH (6284) - mathusee.com
sales@mathusee.com

Beta Student Workbook: Multiple-Digit Addition and Subtraction

©2012 Demme Learning
Published and distributed by Demme Learning

mathusee.com

1-888-854-6284 or +1 717-283-1448 | demmelearning.com
Lancaster, Pennsylvania USA

ISBN 978-1-60826-067-6
Revision Code 1218-C

Printed in the United States of America by The P.A. Hutchison Company
 2 3 4 5 6 7 8 9 10

For information regarding CPSIA on this printed material call: 1-888-854-6284
and provide reference #1218-08132020

Beta β

	LESSON PRACTICE			SYSTEMATIC REVIEW			
	A	B	C	D	E	F	TEST
1 Place Value							
2 Sequencing							
3 Inequalities							
4 Round to 10							
5 Add Multiple Digit							
6 Skip Count 2							
7 Regrouping							
Unit Test 1...........							
8 Skip Count 10, Coins							
9 Skip Count 5, Nickel							
10 Money							
11 Round to 100							
12 Add Money							
13 Column Addition							
14 Measure: Foot							
15 Perimeter							
Unit Test 2...........							
16 1,000s							
17 Round to 1,000s							
18 Multi-Digit Addition							
19 More Multi-Digit Addition							
20 Multi-Digit Subtraction							
21 Time: Minutes							
22 Regrouping							
Unit Test 3...........							
23 Time: Hours							
24 Subtraction: 3 Digits							
25 Ordinal Numbers							
26 Subtraction: 4 Digits							
27 Subtraction: Money							
28 Subtraction: Multi-Digit							
29 Read Gauges							
30 Graphs							
Unit Test 4...........							
Final Test...........							

APPLICATION AND ENRICHMENT PAGES

This edition of the *Beta Student Workbook* includes extra activity pages titled "Application and Enrichment." You will find one enrichment page after the last systematic review page for each lesson. The concepts taught in *Beta* build on what was learned in *Alpha*. Some of the extra activities review and reinforce the addition and subtraction facts. Other activities are intended to do the following:

- Provide enjoyable practice of lesson concepts.
- Stimulate thinking by presenting concepts in different formats.
- Include activities suitable for a wide range of learning styles.
- Enrich learning with additional age-appropriate activities.
- Introduce new concepts that may be useful to students at this level.

The Application and Enrichment pages may be scheduled any time after the student has completed the corresponding lesson. Some activities may be challenging or require a new way of looking at a concept. Do not hesitate to give the student as much help as needed, and remember to have fun.

Check your instruction manual for helpful teaching tips. There are no solutions for the *Beta* Application and Enrichment pages.

Count and write the number. Then say it.

1.

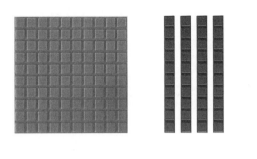

_____ _____ _____

2.

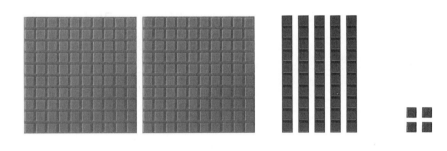

_____ _____ _____

Build and say the number.

3. 28

4. 336

Review your addition facts. These problems review adding 0, 1, and 2.

5. $\begin{array}{r} 3 \\ +\ 1 \\ \hline \end{array}$

6. $\begin{array}{r} 5 \\ +\ 0 \\ \hline \end{array}$

7. $\begin{array}{r} 2 \\ +\ 8 \\ \hline \end{array}$

8. $\begin{array}{r} 0 \\ +\ 7 \\ \hline \end{array}$

9. $\begin{array}{r} 5 \\ +\ 2 \\ \hline \end{array}$

10. $\begin{array}{r} 2 \\ +\ 1 \\ \hline \end{array}$

11. $\begin{array}{r} 1 \\ +\ 5 \\ \hline \end{array}$

12. $\begin{array}{r} 6 \\ +\ 2 \\ \hline \end{array}$

13. 4 + 2 = _____

14. 2 + 9 = _____

15. 7 + 2 = _____

Count and write the number. Then say it.

1.

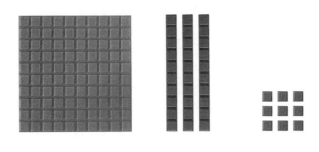

_____ _____ _____

2.

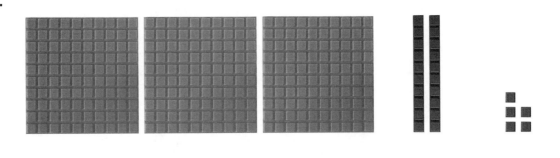

_____   _____ _____

Build and say the number.

3. 31

4. 228

Add. These problems review adding 8 and 9.

5.   9
     $+ 3$

6.   8
     $+ 7$

7.   9
     $+ 5$

8.   8
     $+ 3$

9.   5
     $+ 8$

10.   6
      $+ 9$

11.   9
      $+ 4$

12.   8
      $+ 9$

13. $9 + 7 =$ _____

14. $6 + 8 =$ _____

15. $8 + 4 =$ _____

LESSON PRACTICE

Count and write the number. Then say it.

1.

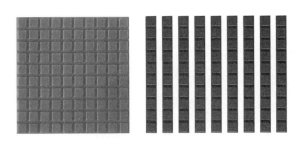

_____   _____   _____

2.

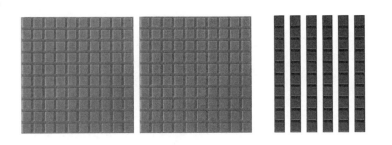

_____   _____   _____

Build and say the number.

3. 49

4. 283

Add. These problems review doubles and doubles plus one.

5.   8
   + 8
   ----

6.   7
   + 7
   ----

7.   7
   + 6
   ----

8.   4
   + 4
   ----

9.   5
   + 5
   ----

10.   6
    + 5
    ----

11.   3
    + 3
    ----

12.   2
    + 2
    ----

13. 6 + 6 = _____

14. 3 + 4 = _____

15. 9 + 9 = _____

Count and write the number. Then say it.

1.

_____   _____   _____

2.

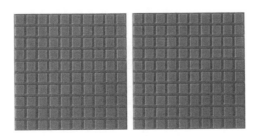

_____   _____   _____

Build and say the number.

3. 72

4. 417

Add. These problems review making 9 and making 10.

5.   8
   + 1

6.   7
   + 3

7.   3
   + 6

8.   1
   + 9

9.   6
   + 4

10.   5
   + 5

11.   7
   + 2

12.   5
   + 4

13. 2 + 8 = _____

14. 3 + 7 = _____

15. 4 + 6 = _____

Count and write the number. Then say it.

1.

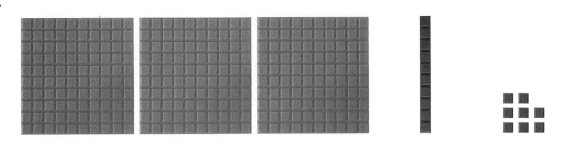

_____   _____   _____

2.

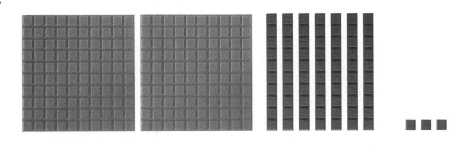

_____   _____   _____

Build and say the number.

3. 50

4. 190

Add. These problems review 4 + 7, 5 + 7, 3 + 5, and other facts.

5.   4
    + 7

6.   7
    + 5

7.   3
    + 5

8.   5
    + 7

9.   5
    + 3

10.   7
    + 4

11.   9
    + 8

12.   5
    + 3

13. 7 + 6 = _____

14. 8 + 2 = _____

15. 1 + 0 = _____

Count and write the number. Then say it.

1.

_____   _____   _____

2.

_____   _____

Build and say the number.

3. 306

4. 222

Add.

5. $\begin{array}{r} 1 \\ + 1 \\ \hline \end{array}$

6. $\begin{array}{r} 7 \\ + 2 \\ \hline \end{array}$

7. $\begin{array}{r} 9 \\ + 5 \\ \hline \end{array}$

8. $\begin{array}{r} 3 \\ + 8 \\ \hline \end{array}$

9. $\begin{array}{r} 6 \\ + 6 \\ \hline \end{array}$

10. $\begin{array}{r} 7 \\ + 8 \\ \hline \end{array}$

11. $\begin{array}{r} 5 \\ + 4 \\ \hline \end{array}$

12. $\begin{array}{r} 3 \\ + 5 \\ \hline \end{array}$

13. 5 + 8 = _____

14. 9 + 4 = _____

15. 3 + 3 = _____

16. 6 + 7 = _____

17. 7 + 5 = _____

18. 4 + 8 = _____

Add.

If the answer is 12, color the space black or gray.
If the answer is 13, leave the space white.
If the answer is 14, color the space dark blue.
If the answer is 15, color the space light blue.

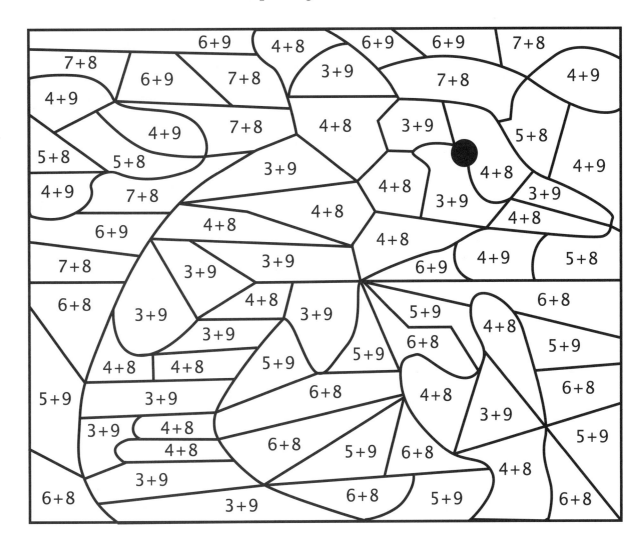

Add.

If the answer is 10, color the space orange.
If the answer is 11, color the space yellow.
If the answer is 12, color the space green.
If the answer is 13, color the space blue.

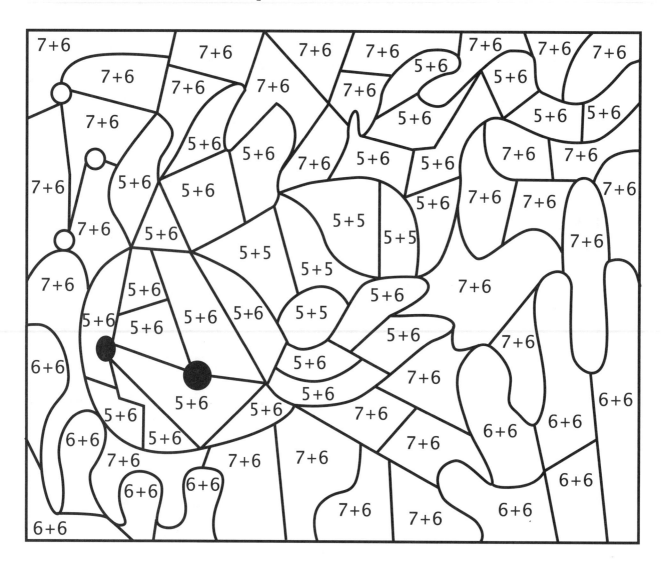

Put the numbers in order from the least to the greatest.

1. 20, 12, 4

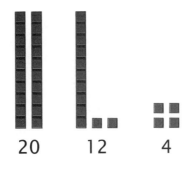

20   12   4

_____, _____, _____

2. 6, 206, 31

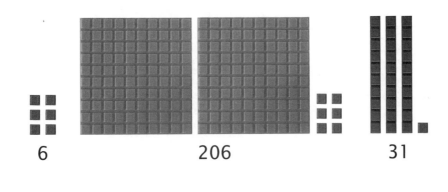

6   206   31

_____, _____, _____

Put the numbers in order from the greatest to the least. Use the blocks to show your answer.

3. 4, 55, 162

_____, _____, _____

4. 16, 3, 136

_____, _____, _____

Fill in the blanks with the correct numbers for each sequence.

5. 8, _____, 10, 11, _____

6. 10, 9, _____, _____, _____

7. 71, _____, _____, 74, _____

Put the numbers in order from the least to the greatest. Use the blocks to show your answer.

1. 16, 5, 8

_____ , _____ , _____

2. 25, 100, 63

_____ , _____ , _____

3. 17, 7, 107

_____ , _____ , _____

4. 200, 89, 11

_____ , _____ , _____

Put the numbers in order from the greatest to the least. Use the blocks to show your answer.

5. 6, 10, 30

_____ , _____ , _____

6. 20, 80, 10

_____ , _____ , _____

7. 9, 245, 61

_____ , _____ , _____

8. 3, 84, 100

_____ , _____ , _____

Fill in the blanks with the correct numbers for each sequence.

9. 6, _____, _____, 3, _____

10. _____, 23, _____, 25, _____

11. 64, _____, _____, _____, 68

12. _____, 16, 15, _____ , _____

Put the numbers in order from the least to the greatest. Use the blocks to show your answer.

1. 100, 50, 6

_____ , _____ , _____

2. 95, 18, 32

_____ , _____ , _____

3. 10, 400, 40

_____ , _____ , _____

4. 243, 65, 16

_____ , _____ , _____

Put the numbers in order from the greatest to the least. Use the blocks to show your answer.

5. 125, 5, 15

_____ , _____ , _____

6. 9, 170, 106

_____ , _____ , _____

7. 22, 99, 48

_____ , _____ , _____

8. 114, 89, 120

_____ , _____ , _____

Fill in the blanks with the correct numbers for each sequence.

9. 29, _____, _____, 26, _____

10. _____, 44, _____, 46, _____

11. 30, _____, _____, _____, 34

12. _____, 8, 7, _____, _____

Put the numbers in order from the least to the greatest.

1. 17, 15, 5

2. 11, 40, 8

_____ , _____ , _____

_____ , _____ , _____

Put the numbers in order from the greatest to the least.

3. 75, 12, 43

4. 150, 200, 110

_____ , _____ , _____

_____ , _____ , _____

Fill in the blanks with the correct numbers for the sequence.

5. 79, _____, _____, 82, _____

Build and say the number.

6. 47

7. 109

Add.

8.   9
     + 9

9.   7
     + 8

10.   4
      + 5

11.   2
      + 3

12. 4 + 8 = _____

13. 5 + 7 = _____

14. 3 + 8 = _____

15. Sam had 121 marbles, and Jeff had 112 marbles. Who had the greater number of marbles?

     _____

16. Riley made four toasted-cheese sandwiches and one peanut butter and jelly sandwich. How many sandwiches did she make in all?

     _____

See Lesson 2 in the Instruction Manual for tips on teaching word problems.

Put the numbers in order from the least to the greatest.

1. 63, 18, 5

2. 100, 400, 200

_____ , _____ , _____

_____ , _____ , _____

Put the numbers in order from the greatest to the least.

3. 44, 14, 56

4. 105, 200, 50

_____ , _____ , _____

_____ , _____ , _____

Fill in the blanks with the correct numbers for the sequence.

5. 9, _____, _____, 6, _____

Build and say the number.

6. 98

7. 276

Add.

8.   6
   + 6

9.   5
   + 9

10.   2
   + 7

11.   4
   + 6

12. $8 + 9 =$ _____

13. $5 + 6 =$ _____

14. $1 + 7 =$ _____

15. Mary picked 7 daisies and 16 roses. Which number is less?

_____

16. Jared counted seven trucks and nine cars going by his house this morning. How many vehicles did he count in all?

_____

Put the numbers in order from the least to the greatest.

1. 39, 19, 99

2. 60, 80, 10

_____ , _____ , _____   _____ , _____ , _____

Put the numbers in order from the greatest to the least.

3. 299, 18, 74

4. 48, 21,180

_____ , _____ , _____   _____ , _____ , _____

Fill in the blanks with the correct numbers for the sequence.

5. _____ , 89, _____ , _____ , 92

Build and say the number.

6. 240

7. 16

Add.

8. $\begin{array}{r} 7 \\ + 7 \\ \hline \end{array}$

9. $\begin{array}{r} 2 \\ + 5 \\ \hline \end{array}$

10. $\begin{array}{r} 3 \\ + 7 \\ \hline \end{array}$

11. $\begin{array}{r} 0 \\ + 4 \\ \hline \end{array}$

12. 8 + 8 = _____

13. 6 + 7 = _____

14. 4 + 9 = _____

15. Emma has 26 school books and 62 story books in her bookcase. Which number is greater?

    _____

16. Kayla lost six pennies and eight dimes. How many coins did she lose in all?

    _____

We can use numbers to make patterns. Connect all the numbers that make ten. Each dot will connect with two other dots. Color the design, if you wish.

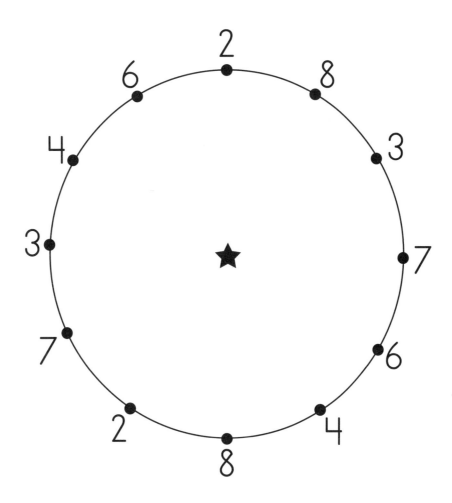

1. Jack saw five lions and five tigers in his living room. How many animals did he see altogether? _____

2. Answer the problem. Make up your own silly word problem. Write it down or tell it to your teacher.

   6 + 4 = _____

Connect all the numbers that make nine. Each dot will connect with two other dots. Color the design, if you wish.

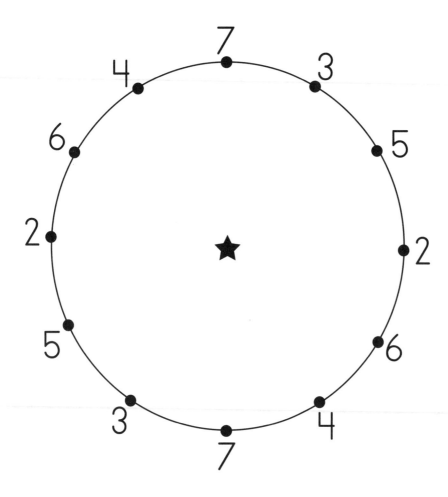

1. Six parrots called me on the telephone yesterday. Three more parrots called me today. How many parrots called me altogether?

   _____

2. Answer the problem. Make up your own silly word problem. Write it down or tell it to your teacher.

   2 + 7 = _____

Compare. Then fill in the oval with <, >, or =.

1. 4 〇 3

2. 5 + 2 〇 2 + 5

3. 4 + 2 〇 3 + 1

4. 11 〇 21

5. 14 〇 7 + 7

6. 2 + 8 〇 6 + 3

7. 63 〇 36

8. 6 + 6 〇 12

9. 6 + 0 〇 7

10. 3 + 9 〇 5 + 5

11. Jennifer's book cost 50 cents, and Nelson's book cost 90 cents. Write an inequality showing whose book cost more. (You can write your answer in two different ways.)

_____

12. Denise did five chores for her mom and four chores for her dad. Michael did five chores for his mom and five chores for his dad. Show who did more chores.

_____

13. Katie read 12 books, and Ethan read 8 books. Show who read fewer books.

_____

14. Chris said that 9 + 5 is the same as 7 + 7. Write an equation showing what Chris was saying.

_____

Compare. Then fill in the oval with <, >, or =.

1. 5 $\bigcirc$ 7

2. 1 + 1 $\bigcirc$ 1 + 3

3. 4 + 6 $\bigcirc$ 2 + 7

4. 13 $\bigcirc$ 10

5. 18 $\bigcirc$ 9 + 8

6. 5 + 6 $\bigcirc$ 8 + 3

7. 27 $\bigcirc$ 72

8. 7 + 7 $\bigcirc$ 15

9. 5 + 1 $\bigcirc$ 6

10. 8 + 4 $\bigcirc$ 4 + 8

11. William had two rabbits and one duck. His brother had three rabbits and two ducks. Write an inequality showing who had more pets.

   _____

12. Jordan got eight words correct on his spelling test, and Alexa got nine words correct. Write an inequality to show who got more words correct.

   _____

13. Ali said that 6 + 4 is the same as 7 + 3. Write an equation showing what Ali was saying.

   _____

14. Write an inequality showing which is greater: 105 or 150.

   _____

Compare. Then fill in the oval with <, >, or =.

1. 9 $\bigcirc$ 14

2. 2 + 3 $\bigcirc$ 2 + 4

3. 7 + 2 $\bigcirc$ 5 + 2

4. 25 $\bigcirc$ 15

5. 16 $\bigcirc$ 8 + 8

6. 9 + 6 $\bigcirc$ 9 + 8

7. 31 $\bigcirc$ 13

8. 7 + 5 $\bigcirc$ 12

9. 3 + 4 $\bigcirc$ 5

10. 9 + 2 $\bigcirc$ 3 + 9

11. It is 10 days until Rachel's birthday and 110 days until Jenny's birthday. Write an inequality showing who has fewer days until her birthday.

_____

12. Mom spent 55 dollars on groceries and 65 dollars on clothes for her family. Write an inequality showing where Mom spent more money.

_____

13. Bria said that 9 + 1 is the same as 1 + 9. Write an equation showing what Bria was saying.

_____

14. Write an inequality showing which is less: 213 or 231.

_____

Compare. Then fill in the oval with <, >, or =.

1. 5 + 8 ◯ 3 + 9

2. 57 ◯ 75

Put the numbers in order from the least to the greatest.

3. 7, 13, 8

4. 90, 50, 70

_____ , _____ , _____

_____ , _____ , _____

Fill in the blanks with the correct numbers for the sequence.

5. 28, _____, _____, 31, _____

Build and say the number.

6. 211

7. 36

Add.

8.   2
   + 9
   ____

9.   5
   + 5
   ____

10.   3
    + 6
    ____

11.   2
    + 8
    ____

12. 6 + 9 = ___

13. 3 + 4 = ___

14. 0 + 6 = ___

15. Write an inequality showing which is greater: 452 or 254.

_____

16. There are five girls and three boys in the Adams family. How many children are there altogether?

_____

Compare. Then fill in the oval with <, >, or =.

1. 2 + 6 ◯ 6 + 2

2. 91 ◯ 19

Put the numbers in order from the greatest to the least.

3. 2, 12, 200

4. 105, 15, 51

_____ , _____ , _____

_____ , _____ , _____

Fill in the blanks with the correct numbers for the sequence.

5. ____, 18, ____, ____, 21

Build and say the number.

6. 104

7. 13

Add.

8. $\begin{array}{r} 4 \\ + 4 \\ \hline \end{array}$

9. $\begin{array}{r} 1 \\ + 6 \\ \hline \end{array}$

10. $\begin{array}{r} 2 \\ + 4 \\ \hline \end{array}$

11. $\begin{array}{r} 0 \\ + 1 \\ \hline \end{array}$

12. 9 + 9 = _____

13. 7 + 8 = _____

14. 4 + 7 = _____

15. Duncan has three brothers and two sisters. Write an inequality showing whether Duncan has more brothers or more sisters.

    _____

16. Alaina walked three miles and rode her bike three miles. How many miles did she travel in all?

    _____

Compare. Then fill in the oval with <, >, or =.

1. $1 + 5 \bigcirc 3 + 4$ 2. $7 + 8 \bigcirc 8 + 7$

Put the numbers in order from the greatest to the least.

3. 63, 3, 300   4. 91, 74, 197

_____, _____, _____   _____, _____, _____

Fill in the blanks with the correct numbers for the sequence.

5. _____, _____, 11, _____, 9

Build and say the number.

6. 135

7. 240

Add.

8.   7
    + 9

9.   2
    + 2

10.   4
   + 9

11.   3
   + 7

12. $1 + 1 = $ _____

13. $5 + 8 = $ _____

14. $7 + 7 = $ _____

15. Sarah said that $9 + 4$ is the same as $7 + 6$. Write an equation showing what Sarah was saying.

_____

16. Kiley read six pages in her book before lunch and eight pages after lunch. How many pages has she read so far?

_____

Fill in the boxes to show how much money Bria and Betsy have. Put < or > in the blank between the boxes to show who has more.

1. Bria had five dollars, and Betsy had seven dollars. How many dollars does each girl have?

   Bria _____ dollars   _____   Betsy _____ dollars

2. Bria lost two dollars. Betsy did not lose any money. How many dollars does each girl have now?

   Bria _____ dollars   _____   Betsy _____ dollars

3. Bria found four dollars. Betsy lost one dollar. How many dollars does each girl have now?

   Bria _____ dollars   _____   Betsy _____ dollars

4. Bria and Betsy each spent three dollars. How many dollars does each girl have now?

   Bria _____ dollars   _____   Betsy _____ dollars

Draw pictures to help you solve a problem with more than one step.

1. Landon picked four apples. Draw the apples in the box.

2. Jack picked five apples. Draw Jack's apples in the box.

3. How many apples did the boys pick in all? _____

4. Mom used seven apples to make a pie. Cross out the apples Mom used. How many apples are left? _____

5. Each of the boys picked one more apple. Draw the apples they picked. Now how many apples are there in all? _____

Answer the questions.

1. Is 44 closer to 40 or 50? _____

2. Is [blocks] closer to [blocks] or [blocks] ? Circle the right answer.

Round to the nearest ten. The first one has been done for you.

3. 25 → _30_

4. 58 → _____

Round each number to the nearest ten. Estimate the answer. The first one has been done for you.

5.   1 8   (2 0)
   + 5 1   (5 0)
            ‾‾‾‾‾‾
            (7 0)

6.   1 3   (   )
   + 1 2   (   )
            ‾‾‾‾‾‾
            (   )

7.   3 1   (   )
   + 1 8   (   )
            ‾‾‾‾‾‾
            (   )

8.   5 1   (   )
   + 4 2   (   )
            ‾‾‾‾‾‾
            (   )

Add across and down. Then add your answers and see if they match. The first one has been done for you.

9.

6	3	9
1	1	2
7	4	11

3	5	
6	1	

10. Erica read 21 books last month and 26 books this month. Estimate how many books she read the last two months.

_____

Answer the questions.

1. Is 38 closer to 30 or 40? _____

2. Is closer to or ? Circle the right answer.

Round to the nearest ten.

3. 62 → _____

4. 85 → _____

Round to the nearest ten. Estimate the answer.

5.   1 6   (   )
  + 2 3   (   )
        (   )

6.   4 5   (   )
  + 3 1   (   )
        (   )

7.   3 4   (   )
  + 2 2   (   )
        (   )

8.   6 1   (   )
  + 1 7   (   )
        (   )

Add across and down. Then add your answers and see if they match.

9.

3	2	
5	2	

10. When Carol went shopping, she found a coat that cost 57 dollars and a dress that cost 32 dollars. Estimate how many dollars it will cost to buy the dress and the coat.

_____

Answer the questions.

1. Is thirty-one closer to thirty or to forty? _____

2. Is seventy-eight closer to seventy or eighty? _____

Round to the nearest ten.

3. 72 → _____

4. 55 → _____

Round to the nearest ten. Estimate the answer.

5.　　27　　(　　)
　　＋32　　(　　)
　　　　　　(　　)

6.　　16　　(　　)
　　＋11　　(　　)
　　　　　　(　　)

7.　　44　　(　　)
　　＋45　　(　　)
　　　　　　(　　)

8.　　23　　(　　)
　　＋36　　(　　)
　　　　　　(　　)

Add across and down. Then add your answers and see if they match.

9.

4	5	
3	1	

10. It rained 32 inches last year and 26 inches this year. Estimate the total amount of rain that fell in both years.

_____

SYSTEMATIC REVIEW

4D

Round to the nearest ten.

1. 81 → _____

2. 49 → _____

Round to the nearest ten. Estimate the answer.

3.
```
    4 6    (   )
  + 2 2    (   )
           (   )
```

4.
```
    1 5    (   )
  + 3 3    (   )
           (   )
```

Add across and down. Then add your answers and see if they match.

5.

6	1	
2	3	

Compare. Then fill in the oval with <, >, or =.

6. 1 + 3 $\bigcirc$ 2 + 2

7. 16 $\bigcirc$ 21

Add.

8.　　4
　　+ 6

9.　　1
　　+ 9

10.　　6
　　+ 9

11.　　5
　　+ 7

12. Kurt worked 15 hours last week and 12 hours this week. Estimate the total time he worked the last two weeks.

_____

13. King John won 9 battles. King Henry won 8 more battles than King John. How many battles did King Henry win?

_____

14. Looking out at Grandpa's bird feeder, Jeremiah saw two robins, two bluebirds, and six sparrows. How many birds did he see altogether? (HINT: First add 2 + 2. Then add 6 to find your answer.)

_____

Round to the nearest ten.

1. 45 → _____

2. 64 → _____

Round to the nearest ten. Estimate the answer.

3.   3 6   (   )
    + 1 3   (   )
    ─────   ─────
              (   )

4.   6 7   (   )
    + 1 2   (   )
    ─────   ─────
              (   )

Add across and down. Then add your answers and see if they match.

5.

2	1	
7	0	

Compare. Then fill in the oval with <, >, or =.

6. 5 + 9 ⬭ 6 + 7

7. 99 ⬭ 105

Add.

8.　　3
　　+ 9

9.　　1
　　+ 8

10.　　5
　　+ 5

11.　　2
　　+ 7

12. Casey ordered three chocolate and three strawberry ice cream cones for her family. How many cones did she order?

_____

13. Justin likes to fish. He dug up 17 earthworms on Monday and 32 earthworms on Wednesday. Estimate how many worms he has for his next fishing trip.

_____

14. Last week Isaac's Bagel Shop sold 279 bagels. This week it sold 410 bagels. Write an inequality showing which is the greater number of bagels.

_____

Round to the nearest ten.

1. 91 → _____

2. 75 → _____

Round to the nearest ten. Estimate the answer.

3.   5 4   ( )
      + 3 3   ( )
                 ( )

4.   4 8   ( )
      + 2 1   ( )
                 ( )

Add across and down. Then add your answers and see if they match.

5.

3	3	
2	1	

Compare. Then fill in the oval with <, >, or =.

6. 9 + 9 $\bigcirc$ 8 + 8

7. 198 $\bigcirc$ 201

Add.

8.   9
    + 5

9.   2
    + 6

10.   4
    + 3

11.   3
    + 8

12. Haley received nine cards on her birthday. The next day she got two more. How many cards did Haley get in all?

_____

13. A salesman sold 53 radios last month and 36 radios this month. Estimate how many he sold in the last two months.

_____

14. Write an inequality showing which is less: 45 or 405.

_____

Start at 1 and count to 50. Connect the dots to finish the picture.

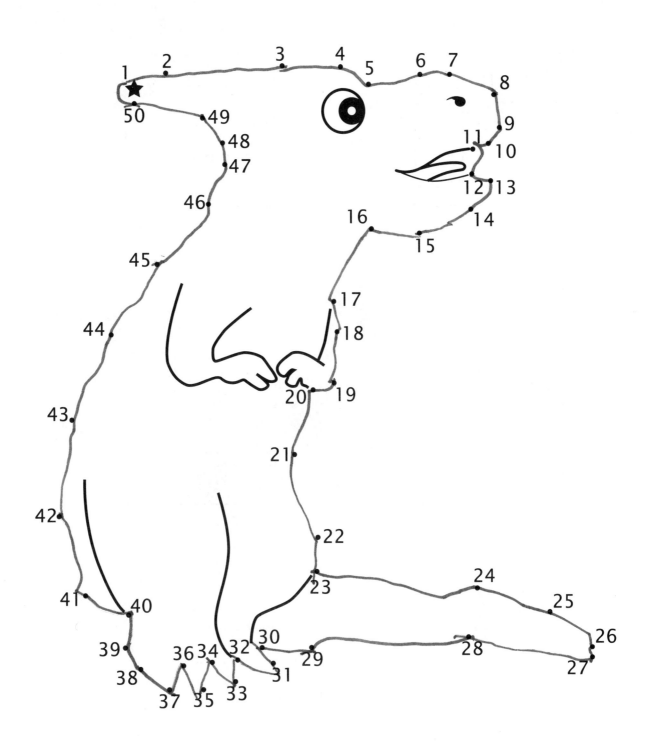

Here are some more addition squares. Add across and down. All of these squares help you practice adding tens.

20	10	
30	30	

10	10	
20	30	

50	10	
10	20	

40	50	
0	10	

Build and write using place-value notation. The first one has been done for you.

1.

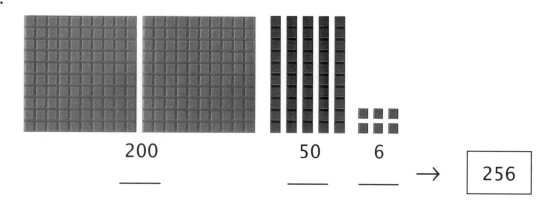

200 _____   50 _____   6 _____ → 256

2.

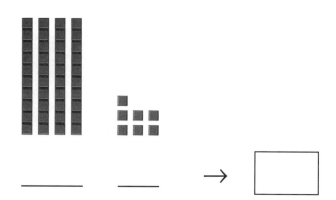

_____   _____ → [ ]

Write using place-value notation.

3. 143 = _____ + _____ + _____

4. 65 = _____ + _____

Add using place-value notation. The first one has been done for you.

5. $\begin{array}{r} 24 \\ +35 \\ \hline 59 \end{array}$ $\rightarrow$ $\begin{array}{r} 20+4 \\ 30+5 \\ \hline 50+9 \end{array}$

6. $\begin{array}{r} 62 \\ +17 \\ \hline \end{array}$ $\rightarrow$ $\begin{array}{r} 60+2 \\ 10+7 \\ \hline \end{array}$

7. $\begin{array}{r} 22 \\ +23 \\ \hline \end{array}$ $\rightarrow$ $\begin{array}{r} 20+2 \\ 20+3 \\ \hline \end{array}$

8. $\begin{array}{r} 141 \\ +128 \\ \hline \end{array}$ $\rightarrow$ $\begin{array}{r} 100+40+1 \\ 100+20+8 \\ \hline \end{array}$

9. There are 24 houses on Green Street and 33 houses on Long Lane. How many houses are there altogether?

   _____

10. Austin spent 16 dollars at the hardware store and 12 dollars at the candy store. How much did he spend in all?

   _____

Build and write using place-value notation.

1.

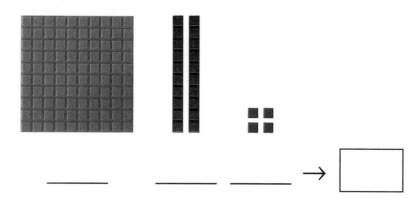

_____   _____   _____   → ☐

2.

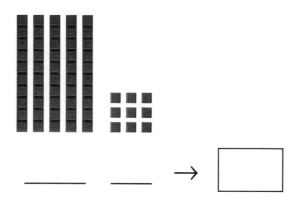

_____   _____   → ☐

Write using place-value notation.

3. 365 = _____ + _____ + _____

4. 41 = _____ + _____

Add using place-value notation.

5.   6 4 → 6 0 + 4
       + 1 5 → 1 0 + 5

6.   1 3 → 1 0 + 3
      + 1 3 → 1 0 + 3

7.   5 7 → 5 0 + 7
      + 1 1 → 1 0 + 1

8.   3 3 1 → 3 0 0 + 3 0 + 1
      +   2 4 →   2 0 + 4

9. Mom spent 133 dollars at the grocery store and 10 dollars at the pet store. How much money did Mom spend in all?

_____

10. Twenty-three blackbirds flew over our home yesterday, and twenty-three more flew over today. How many blackbirds flew over in two days?

_____

Build and write using place-value notation.

1.

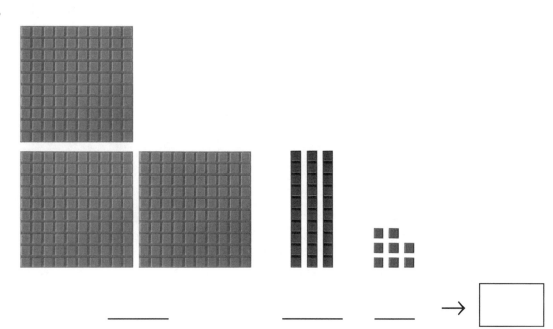

_____   _____ _____ → ☐

2.

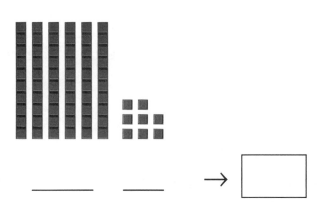

_____ _____ → ☐

Write using place-value notation.

3. 172 = _____ + _____ + _____

4. 27 = _____ + _____

Add using place–value notation.

5.   22 → 20 + 2
     +22 → 20 + 2

6.   55 → 50 + 5
     +41 → 40 + 1

7.   89 → 80 + 9
     +10 → 10 + 0

8.   236 → 200 + 30 + 6
     +212 → 200 + 10 + 2

9. There are 12 students in our class and 36 students in the other class. How many students would there be if the two classes were combined?

_____

10. I found 41 cents in my dresser drawer and 46 cents in my desk. How much money did I find altogether?

_____

Write using place-value notation.

1. 531 = _____ + _____ + _____

2. 18 = _____ + _____

Add using place-value notation.

3.
```
  3 2  →  3 0 + 2
+ 1 1  →  1 0 + 1
```

4.
```
  1 0 1  →  1 0 0 + 0 0 + 1
+ 3 2 1  →  3 0 0 + 2 0 + 1
```

Round to the nearest ten and estimate the answer. Then find the exact answer. The first one has been done for you.

5.
```
  6 7    (7 0)
+ 2 2    (2 0)
  8 9    (9 0)
```

6.
```
  4 3    (   )
+ 2 1    (   )
         (   )
```

Compare. Then fill in the oval with <, >, or =.

7. 3 + 4 ⬭ 6 + 6

8. 71 ⬭ 17

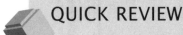

QUICK REVIEW

Solving for the unknown is a good way to review addition and get ready for subtraction.*

Example 1

$$4 + A = 9 \qquad \overset{5}{4 + A = 9}$$

Think, "Four plus what equals nine?"

The answer is five.

Write the answer above the letter that stands for the unknown in the equation.

You may use the blocks by laying the four bar next to the nine bar and finding what bar is needed to make up the length. The letter for the unknown may replace any number in a problem.

Solve for the unknown.

9. $6 + X = 12$

10. $7 + R = 11$

11. Sue read five books about stars. She also read some books about elephants. If Sue read 7 books altogether, how many were about elephants? $B + 5 = 7$

12. One hundred twenty geese and two hundred fifty-four swans flew overhead. How many birds did I see?

_____

*Subtraction facts are systematically reviewed in Lessons 11–19 in preparation for multiple-digit subtraction. If your student needs more review and practice of addition or subtraction facts, go to mathusee.com, where you can find online drill and downloadable worksheets.

Write using place-value notation.

1. 114 = _____ + _____ + _____

2. 39 = _____ + _____

Add using place-value notation.

3. $\begin{array}{r} 17 \\ +22 \end{array}$ → $\begin{array}{r} 10 + 7 \\ 20 + 2 \end{array}$

4. $\begin{array}{r} 242 \\ +111 \end{array}$ → $\begin{array}{r} 200 + 40 + 2 \\ 100 + 10 + 1 \end{array}$

Round to the nearest ten and estimate. Then find the exact answer.

5. $\begin{array}{r} 68 \\ +11 \end{array}$ ( )
   ( )
   ( )

6. $\begin{array}{r} 51 \\ +34 \end{array}$ ( )
   ( )
   ( )

Add across and down. Then add your answers and see if they match.

7.

5	6	
7	1	

Compare. Then fill in the oval with <, >, or =.

8. $7 + 8 \bigcirc 8 + 7$ 9. $103 \bigcirc 331$

Solve for the unknown.

10. $7 + X = 10$ 11. $9 + R = 16$

12. Bill saw 5 spiders on Tuesday. He saw more spiders on Wednesday. If Bill saw 9 spiders in all, how many did he see on Wednesday? Solve for the unknown.

13. Tom got 18 questions right on his first test and 21 right on the second. Estimate the total number he got right.

_____

14. We drove 212 miles this morning and 362 miles this afternoon. How many miles did we drive today?

_____

Write using place-value notation.

1. 314 = ＿＿ + ＿＿ + ＿＿

2. 72 = ＿＿ + ＿＿

Add using place-value notation.

3. $\begin{array}{r} 1\,1 \\ +\,8\,8 \\ \hline \end{array}$ → $\begin{array}{r} 1\,0+1 \\ 8\,0+8 \\ \hline \end{array}$

4. $\begin{array}{r} 1\,7\,6 \\ +\,3\,2\,3 \\ \hline \end{array}$ → $\begin{array}{r} 1\,0\,0+7\,0+6 \\ 3\,0\,0+2\,0+3 \\ \hline \end{array}$

Round to the nearest ten and estimate. Then find the exact answer.

5. $\begin{array}{r} 7\,2 \\ +\,1\,5 \\ \hline \end{array}$ ( )
( )
( )

6. $\begin{array}{r} 6\,1 \\ +\,1\,7 \\ \hline \end{array}$ ( )
( )
( )

Add across and down. Then add your answers and see if they match.

7.

2	8	
6	3	

Compare. Then fill in the oval with <, >, or =.

8. 2 + 4 ⬭ 4 + 2   9. 86 ⬭ 68

Solve for the unknown.

10. 8 + T = 13   11. 4 + Q = 5

12. Dad gave Nate 9 peanuts. Mom gave him some more peanuts. If Nate had 11 peanuts in all, how many had Mom given him? Solve for the unknown.

13. If Farmer John had 665 cows and 133 calves in the barn, how many animals did he have in the barn?

    _____

14. Write an inequality showing which is greater: 21 + 21 or 12 + 12.

    _____

Answer the questions and fill in the crossword puzzle.

Across

3. The numbers we add are called _____.

5. Which one is greater, two or five? _____

6. The house next to the units house on Decimal Street is the _____ house.

Down

1. Next to the tens is the big red _____ castle.

2. How many people can live in each house on Decimal Street? _____

4. The answer to an addition problem is the _____.

To the parent: The names for the parts of an addition problem are introduced in Lesson 5 in the Instruction Manual. Please give the student as much help as is needed with this enrichment page.

Color by number.

If the answer is 29, color the space blue.
If the answer is 39, color the space yellow.
If the answer is 55, color the space black or gray.
If the answer is 76, color the space purple.

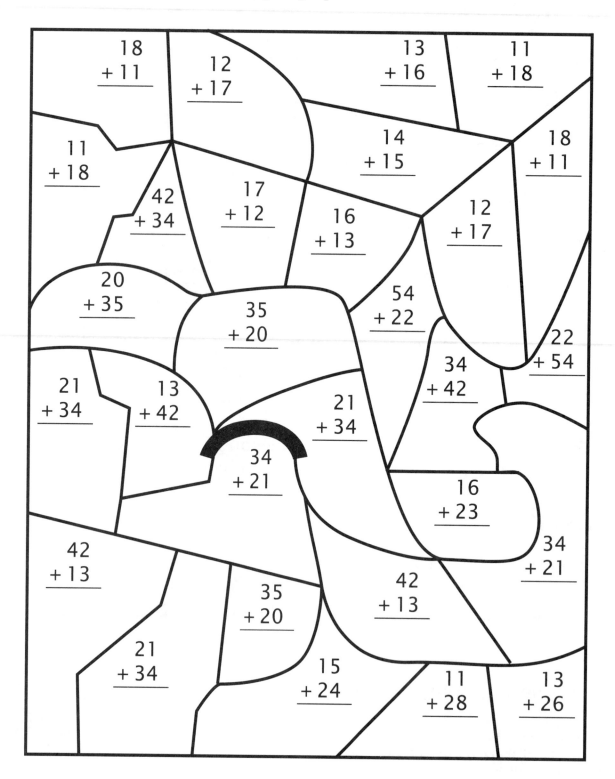

Skip count and write the numbers on the lines in the boxes. Then write the numbers in the spaces underneath. The numbers that you say when you skip count by two are called even numbers.

1.

	2
	___
	6
	___
	___
	12
	___
	___
	___
	20

2.

	___
	4
	___
	8
	10
	___
	___
	___
	___
	___

3. 2, 4, _____, _____, _____, _____, _____, _____, _____, _____

Skip count by two to find each answer.

4. How many cookies? _____

5. How many Xs? _____

6. There are nine children at the party, all wearing their best shoes. How many shoes are there in all?

    _____

6B

Skip count and write the numbers on the lines in the boxes. Then write the numbers in the spaces underneath.

1.

	2
	___
	___
	8
	___
	___
	___
	16
	18
	___

2.

	___
	4
	6
	___
	10
	___
	___
	___
	___
	___

3. 2, _____, _____, 8, _____, _____, _____, _____, _____, _____

Skip count by two to find each answer.

4. How many cookies? _____

5. How many Xs? _____

6. Five elephants are at the zoo. How many great big elephant ears can you see?

_____

6C

Skip count and write the numbers on the lines in the boxes. Then write the numbers in the spaces underneath.

1.

	___
	4
	___
	___
	___
	12
	___
	___
	___
	20

2.

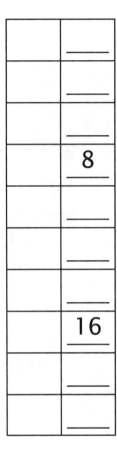

	___
	___
	___
	8
	___
	___
	16
	___
	___

3. 2, _____, _____, _____, _____, _____, _____, _____, 18, _____

Skip count by two to find each answer.

4. How many snowflakes? _____

5. Five basketball players lifted both arms to get the rebound. Skip count to find how many arms were in the air.

    _____

6. Ava read two books a week. After three weeks, how many books had Ava read?

    _____

Skip count by two and write the numbers.

1. 2, _____, _____, 8, _____, _____, _____, _____, _____, _____

Add using place-value notation.

2.   11 → 10 + 1
     + 11 → 10 + 1

3.   171 → 100 + 70 + 1
     + 118 → 100 + 10 + 8

Round to the nearest ten and estimate. Then find the exact answer.

4.   13   ( )
    + 15 ( )
           ( )

5.   32   ( )
    + 41 ( )
           ( )

Compare. Then fill in the oval with <, >, or =.

6. 5 + 3 ◯ 2 + 4

7. 108 ◯ 801

Solve for the unknown.

8. 5 + X = 8

9. 8 + R = 15

10. 9 + B = 13

11. David remembered to brush his teeth two times a day. Skip count to find how many times he brushed his teeth in one week (seven days).

    _____

12. Sam smiled at his brother seven times, and then he smiled at Debbie two times. Since he was a very happy baby, he smiled three more times after that. How many times did Sam smile altogether?

    _____

Skip count by two and write the numbers.

1. 2, _____, _____, _____, _____, _____, _____, _____, _____, 20

Add using place-value notation.

2.   3 4 → 3 0 + 4
   +   4 → 0 0 + 4
                  ———————

3.   3 1 9 → 3 0 0 + 1 0 + 9
  +3 8 0 → 3 0 0 + 8 0 + 0
                  ————————————

Round to the nearest ten and estimate. Then find the exact answer.

4.   4 8   (   )
   + 2 1 (   )
            —————
            (   )

5.   5 4   (   )
   + 1 2 (   )
            —————
            (   )

Compare. Then fill in the oval with <, >, or =.

6. 8 + 2 $\bigcirc$ 5 + 3

7. 295 $\bigcirc$ 592

Solve for the unknown.

8. $3 + A = 3$ 9. $5 + F = 8$

10. $4 + H = 9$

11. Robin spent 126 dollars for groceries the first week and 132 dollars the second week. How much did she spend for groceries in two weeks?

    _____

12. Each of the six deer gave birth to twin fawns. Skip count to find how many fawns were born.

    _____

Skip count by two and write the numbers.

1. ____, 4, ____, ____, ____, ____, ____, ____, 18, ____

Add using place-value notation.

2.   6 0 → 6 0 + 0
      +   3 → 0 0 + 3

3.   4 9 2 → 4 0 0 + 9 0 + 2
      + 2 0 4 → 2 0 0 + 0 0 + 4

Round to the nearest ten and estimate. Then find the exact answer.

4.   1 1   (   )   5.   5 3   (   )
      + 3 3   (   )   + 2 5   (   )
                 (   )   (   )

Compare. Then fill in the oval with <, >, or =.

6. 6 + 1 ◯ 1 + 6   7. 21 + 21 ◯ 12 + 12

Solve for the unknown.

8. $7 + Q = 9$

Fill in the blanks with equal addends to make the doubles. The answer to a doubles addition problem is always an even number. The first one has been done for you.

9. $\underline{6} + \underline{6} = 12$

10. $\underline{} + \underline{} = 14$

11. Rachel grew two inches every year for four years. Skip count to find how many inches she grew in all.

_____

12. Evan found 11 pennies in his pocket and 6 pennies under the bed. How many pennies did he find?

_____

APPLICATION AND ENRICHMENT

Start at the star and skip count by two.

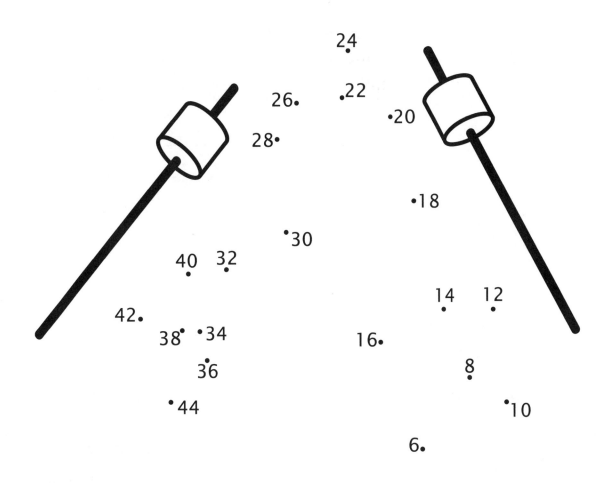

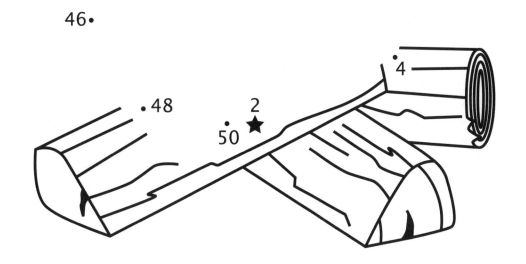

Use pictures to help you solve word problems with more than one step. Draw more pictures to add. Make an X on pictures to subtract. You may use the blocks as well.

1. Rachel found six sea shells. She lost two of them. Rena found five sea shells and lost three of them. Then the girls found four more shells. How many shells do the girls have altogether?

2. John made four snowballs and threw two of them at his brother. He made four more snowballs and threw two of them at his sister. Then he made six more snowballs. Mom called him in to dinner, so he left all his snowballs in a pile. How many snowballs are left in the pile?

Add using place-value notation and regrouping. The first one has been done for you.

$$
\begin{array}{r}
{\scriptstyle 1} \\
34 \\
+\,28 \\
\hline
62
\end{array}
$$

1.   3 4 → 3 0 + 4
    + 2 8 → 2 0 + 8
      6 2   6 0 + 2

2.   4 6 → 4 0 + 6
    + 3 5 → 3 0 + 5

3.   7 3 → 7 0 + 3
    + 1 8 → 1 0 + 8

4.   3 5 → 3 0 + 5
    + 3 7 → 3 0 + 7

5.   2 6 → 2 0 + 6
    + 5 5 → 5 0 + 5

6.   3 8 → 3 0 + 8
    + 4 4 → 4 0 + 4

Regroup and add.

7. $\begin{array}{r} 4\ 7 \\ +\ 2\ 5 \\ \hline \end{array}$
8. $\begin{array}{r} 6\ 6 \\ +\ 3\ 3 \\ \hline \end{array}$

9. $\begin{array}{r} 7\ 5 \\ +\ 1\ 6 \\ \hline \end{array}$

10. Thomas has two sacks of marbles. One sack has 59 marbles, and the other sack has 37 marbles. How many marbles does Thomas have?

   _____

LESSON PRACTICE

Add using place-value notation and regrouping.

1. $\begin{array}{r} 35 \\ +25 \end{array}$ → $\begin{array}{r} 30 + 5 \\ 20 + 5 \end{array}$

2. $\begin{array}{r} 59 \\ +24 \end{array}$ → $\begin{array}{r} 50 + 9 \\ 20 + 4 \end{array}$

3. $\begin{array}{r} 45 \\ +28 \end{array}$ → $\begin{array}{r} 40 + 5 \\ 20 + 8 \end{array}$

4. $\begin{array}{r} 76 \\ +15 \end{array}$ → $\begin{array}{r} 70 + 6 \\ 10 + 5 \end{array}$

5. $\begin{array}{r} 24 \\ +9 \end{array}$ → $\begin{array}{r} 20 + 4 \\ + 9 \end{array}$

6. $\begin{array}{r} 25 \\ +39 \end{array}$ → $\begin{array}{r} 20 + 5 \\ 30 + 9 \end{array}$

Regroup and add.

7.
```
   2 4
 + 4 8
```

8.
```
   6 7
 +   8
```

9.
```
   5 6
 + 2 4
```

10. Richard and Sarah baked cookies. Sarah baked 26 cookies, and Richard baked 38 cookies. How many cookies did they bake altogether?

   _____

Add using place-value notation and regrouping.

1.　　25 → 20 + 5
　　+ 65 → 60 + 5

2.　　34 → 30 + 4
　　+ 45 → 40 + 5

3.　　72 → 70 + 2
　　+ 18 → 10 + 8

4.　　68 → 60 + 8
　　+　8 → ＿＿ + 8

5.　　34 → 30 + 4
　　+ 46 → 40 + 6

6.　　36 → 30 + 6
　　+ 45 → 40 + 5

Regroup and add.

7.
```
  4 8
+   7
```

8.
```
  2 4
+ 6 6
```

9.
```
  1 8
+ 5 3
```

10. For his birthday, Michael got 16 new baseball cards. He already had 8 cards. How many cards does he have now?

_____

SYSTEMATIC REVIEW

Add. Regroup if needed.

1. $\begin{array}{r} 4\ 9 \\ +\ 4\ 5 \\ \hline \end{array}$

2. $\begin{array}{r} 3\ 6 \\ +\ 3\ 6 \\ \hline \end{array}$

3. $\begin{array}{r} 6\ 8 \\ +\ 2\ 5 \\ \hline \end{array}$

4. $\begin{array}{r} 5\ 5 \\ +\ 1\ 4 \\ \hline \end{array}$

5. $\begin{array}{r} 7\ 7 \\ +\ \ \ 7 \\ \hline \end{array}$

6. $\begin{array}{r} 9\ 5 \\ +\ \ \ 3 \\ \hline \end{array}$

Skip count by two and write the numbers.

7. ____ , 4 , _____ , _____ , 10, _____ , _____ , _____ , _____ , _____

Compare. Then fill in the oval with <, >, or =.

8. 9 + 3 $\bigcirc$ 7 + 6

9. 113 $\bigcirc$ 103

Solve for the unknown.

10. 8 + B = 14   11. 6 + G = 10

12. 9 + K = 15

Add across and down. Then add your answers and see if they match.

13.

1	4	
7	3	

14. The clown at Gabriel's party told two jokes every minute. Skip count to find how many jokes the clown told in six minutes.

_____

15. Ian slid down the water slide 14 times, and Jamie slid down 17 times. How many trips down the water slide did the boys make altogether?

_____

Add. Regroup if needed.

1.   12
   + 13

2.   37
   + 28

3.   63
   + 36

4.   22
   + 48

5.   82
   + 9

6.   52
   + 6

Skip count by two and write the numbers.

7. _____, _____, 6, _____, _____, _____, _____, _____, _____, 20

Round to the nearest ten.

8. 16 → _____

9. 23 → _____

Solve for the unknown.

10. $9 + B = 14$

11. $5 + G = 6$

12. $7 + K = 13$

Add across and down. Then add your answers and see if they match.

13.

2	6	
5	8	

14. Katherine thinks that 8 + 3 is larger than 6 + 2. Write an inequality showing what Katherine thinks.

   _____

15. Mom spent 125 dollars at the grocery store and 172 dollars to have the car fixed. How much money did she spend?

   _____

Add. Regroup if needed.

1. $\begin{array}{r} 4\ 3 \\ +\ 2\ 6 \\ \hline \end{array}$

2. $\begin{array}{r} 1\ 9 \\ +\ 1\ 8 \\ \hline \end{array}$

3. $\begin{array}{r} 6\ 2 \\ +\ 3\ 2 \\ \hline \end{array}$

4. $\begin{array}{r} 1\ 5 \\ +\ 3\ 6 \\ \hline \end{array}$

5. $\begin{array}{r} 2\ 9 \\ +\ \ 8 \\ \hline \end{array}$

6. $\begin{array}{r} 7\ 9 \\ +\ \ 6 \\ \hline \end{array}$

Skip count by two and write the numbers.

7. 2, _____, _____, _____, _____, _____, _____, _____, _____, _____

Round to the nearest ten.

8. 48 → _____

9. 65 → _____

Solve for the unknown.

10. $7 + X = 12$ 11. $8 + W = 11$

12. $9 + A = 17$

Add across and down. Then add your answers and see if they match.

13.

7	8	
6	5	

14. Krista wrapped eight packages for Christmas. She put two bows on each package. Skip count to find how many bows she used.

_____

15. Chris counted 28 cows during the car trip, and his sister Kelly counted 13 horses. Estimate how many large animals they saw during the trip. Then find the exact answer.

_____

Color by number.

If the answer is 33, color the space yellow.
If the answer is 41, color the space green.
If the answer is 87, color the space red.

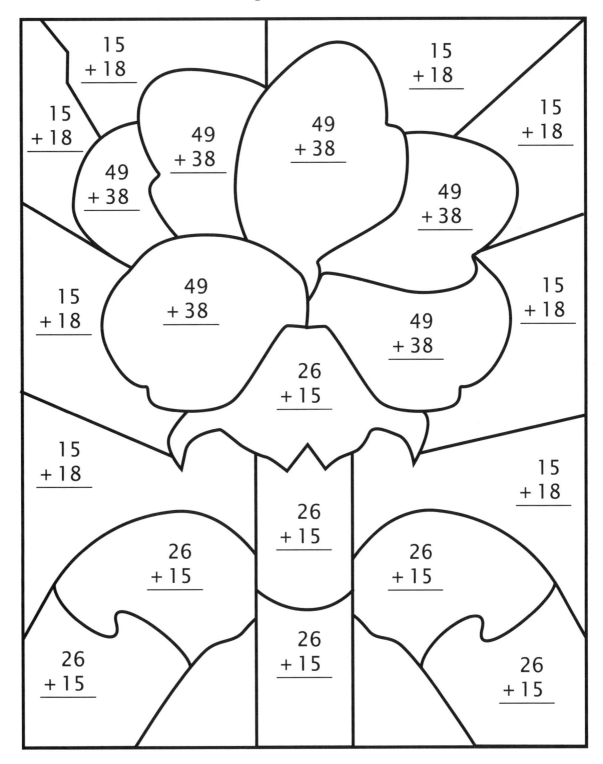

Help to write the word problems. Write your answers in the boxes.
Fill in the blanks with kinds of pets.

1. Riley had 25 _____ and 36 _____ in her pet

    shelter. She found 12 _____ .

    How many pets does she have now?

Fill in the blanks with kinds of sandwiches.

2. Kia made 17 _____ sandwiches and 29 _____

    sandwiches for the picnic. Dylan made 13 more

    _____ sandwiches.

    How many sandwiches were made altogether?

Fill in the blanks with kinds of books.

3. In the bookstore, Jacob counted 16 books about _____

    and 47 books about _____ . Then he counted 7 books

    about _____ .

    How many books did Jacob count in all?

Write the correct numbers in the boxes with lines.

1.

									10
									___
									30
									___
									___
									___
									___
									___
									___
									100

Skip count by 10 and write the numbers.

2. 10, _____, _____, _____, _____, _____, _____, _____, _____, _____

Skip count by 10 to find how many cents.

3. = _____ ¢

4. = _____ ¢

5. Brad has five dimes in his pocket. Skip count to find how many cents he has.

   _____

6. Austin would like to buy 30 little toy animals for his farm. If there are 10 animals in a package, how many packages does he need to buy? Skip count to find the answer.

   _____

8B

Write the correct numbers in the boxes with lines.

1.

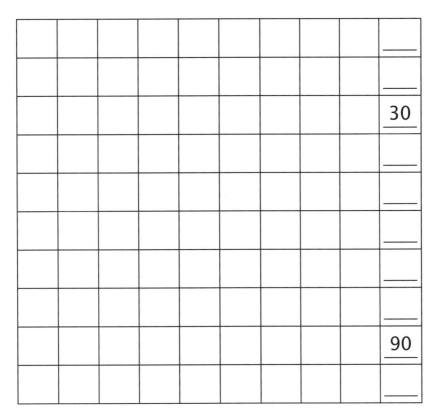

Skip count by 10 and write the numbers.

2. ____, 20, ____, ____, 50, ____, ____, ____, ____, ____

Skip count by 10 to find how many cents.

3.  = ____ ¢

4.  = ____ ¢

5. Each ride at the carnival cost 10¢. How much did I spend if I went on eight rides?

_____

6. Riley has 10 dimes. If she turned her dimes in for pennies, how many pennies would she get?

_____

LESSON PRACTICE

Write the correct numbers in the boxes with lines.

1.

								___
								___
								___
								40
								___
								___
								___
								___
								___
								100

Skip count by 10 and write the numbers.

2. 10, ____, ____, ____, ____, ____, ____, ____, ____, ____

Skip count by 10 to find how many cents.

3.  = ____ ¢

4. = ____ ¢

5. A candy bar costs 30¢. How many pennies would be needed to buy it?

   _____

6. Joseph made ten dollars a day for six days. Skip count to find how much money he has.

   _____

Skip count by 10 and 2. Write the numbers.

1. ____, 20, ____, 40, ____, ____, ____, ____, ____, ____

2. 2, ____, ____, ____, ____, 12, ____, ____, ____, ____

Add. Regroup if needed.

3.   1 1
  + 3 2

4.   6 4
  + 2 6

5.   5 5
  + 1 7

Add. These do not need regrouping.

6.   3 4 1
  + 1 1 1

7.   6 2 9
  + 2 5 0

8.   1 6 5
  + 5 2 2

Solve for the unknown.

9. $6 + X = 7$

10. $6 + R = 8$

11. $3 + D = 3$

12. Jonathan has two dimes. How many pennies would it take to make the same amount of money?

    _____

13. Rachel spent 19 dollars on a new sweater and 36 dollars on a new pair of boots. Estimate how much she spent. Then find the exact amount.

    _____

14. James spotted 41 birds yesterday and 48 birds today. How many birds has he seen in the last two days?

    _____

15. Which is more, 95 pennies or 9 dimes?

    _____

Skip count by 2 and 10. Write the numbers.

1. ____, 4, ____, ____, ____, ____, ____, ____, ____, 20

2. ____, ____, ____, ____, 50, ____, ____, ____, 90, ____

Add. Regroup if needed.

3.
$$\begin{array}{r} 1\,5 \\ +\ 6\,4 \\ \hline \end{array}$$

4.
$$\begin{array}{r} 1\,3 \\ +\ 2\,8 \\ \hline \end{array}$$

5.
$$\begin{array}{r} 4\,4 \\ +\ 4\,6 \\ \hline \end{array}$$

Add. These do not need regrouping.

6.
$$\begin{array}{r} 1\,7\,5 \\ +\,1\,1\,4 \\ \hline \end{array}$$

7.
$$\begin{array}{r} 7\,3\,2 \\ +\,1\,5\,6 \\ \hline \end{array}$$

8.
$$\begin{array}{r} 2\,4\,4 \\ +\,2\,3\,4 \\ \hline \end{array}$$

Solve for the unknown.

9. 4 + A = 6

10. 6 + X = 10

Fill in the blanks with equal addends to make the double. The sum is an even number.

11. _____ + _____ = 18

12. When Jeff had finished playing, he put away six toys. Then he put away two more. His mother told him to put away three more toys. How many toys has Jeff put away?

_____

13. Jane had two dimes, and her mother gave her five more. How many cents does Jane have?

_____

14. Aunt Lillian is making gifts for her 12 nieces and nephews. She has finished 8 gifts. How many more gifts does she need to make so that everyone has one gift? (Solve for the unknown.)

_____

15. A family of beavers gnawed down 242 trees one day and 157 trees the next day. How many trees have the beavers gnawed down?

_____

Skip count by 10 and 2. Write the numbers.

1. ____, ____, 30, ____, ____, ____, ____, ____, ____, ____

2. ____, ____, 6, ____, ____, ____, ____, ____, ____, ____

Add. Regroup if needed.

3.   2 5
    + 2 5

4.   1 9
    + 4

5.   3 8
    + 1 5

Add. These do not need regrouping.

6.   4 3 0
    + 2 2 3

7.   8 0 5
    + 1 9 2

8.   3 1 7
    + 6 5 1

Solve for the unknown.

9. $1 + Q = 8$

10. $9 + Y = 17$

11. $4 + T = 5$

12. Joseph cut 56 pieces of red paper and 39 pieces of green paper to make a paper chain for his Christmas tree. How many pieces did he cut?

_____

13. John's mules ate two bales of hay every day. Skip count to find how many bales the mules ate in three days.

_____

14. Zarah has eight pennies, and Chance has eight dimes. Who has more money?

_____

15. Sue worked two hours on Wednesday, two hours on Thursday, and five hours on Friday. How many hours did she work in all?

_____

APPLICATION AND ENRICHMENT

Skip count by 10 to 100.

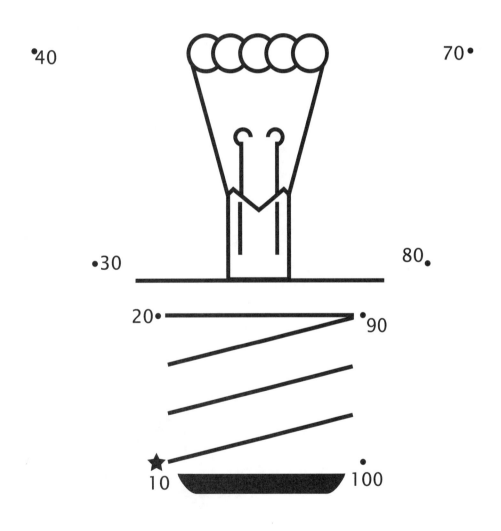

50• •60

•40 70•

•30 80•

20• •90

★
10 •100

Match each thing with the money you need to buy it.

10¢

40¢

80¢

20¢

Write the correct numbers in the boxes with lines.

1.

				___
				10
				15
				___
				___
				___
				___
				___
				___
				50

2.

				5
				___
				___
				___
				___
				___
				___
				___
				___
				___
				___

Skip count by five and write the numbers.

3. 5, _____, _____, _____, _____, _____, _____, _____, _____, _____

Skip count by five to find how many cents.

4.  = _____ ¢

5. Each house has five rooms. Skip count to find how many
    rooms there are in all.

    _____

6. Shelly's mom gave her seven nickels. How many cents does
    she have?

    _____

7. Danica ate five jelly beans after lunch every day. How many
    jelly beans did she eat in nine days?

    _____

Write the correct numbers in the boxes with lines.

1.

				5
				__
				__
				20
				__
				__
				__
				45
				__

2.

				__
				__
				15
				__
				__
				__
				__
				__
				__

Skip count by five and write the numbers.

3. ____, ____, ____, ____, 25, ____, ____, 40, ____, ____

Skip count by five to find how many cents.

4.  = ____ ¢

5. Each shape (pentagon) has five sides. How many sides are there in all?

_____

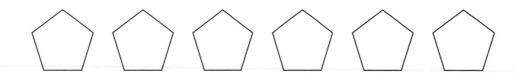

6. Oranges cost a nickel each. How much will 9 oranges cost?

_____

7. Dayna's favorite CD had five songs on it. If she listened to it eight times, how many songs did she hear?

_____

LESSON PRACTICE

Write the correct numbers in the boxes with lines.

1.

				_____
				_____
				_____
				_____
				25
				_____
				_____
				40
				_____
				_____

2.

				_____
				10
				_____
				_____
				_____
				35
				_____
				_____
				_____

Skip count by five and write the numbers.

3. 5, ____, ____, ____, ____, ____, ____, ____, ____, 50

Skip count by five to find how many cents.

4.  = ____ ¢

5. Each flower has five petals. How many petals are there in all?

_____

6. Sue has three nickels, and Bob has two dimes. Which person has more money?

_____

7. Dad gave Chris four nickels to spend. How many cents did Dad give Chris?

_____

Skip count by five and ten. Write the numbers.

1. ____, ____, 15, ____, 25, ____, ____, ____, ____, ____

2. ____, ____, ____, ____, 50, ____, 70, ____, ____, ____

Add. Regroup if needed.

3.
$$\begin{array}{r} 63 \\ +7 \\ \hline \end{array}$$

4.
$$\begin{array}{r} 24 \\ +48 \\ \hline \end{array}$$

5.
$$\begin{array}{r} 15 \\ +44 \\ \hline \end{array}$$

Add. These do not need regrouping.

6.
$$\begin{array}{r} 412 \\ +216 \\ \hline \end{array}$$

7.
$$\begin{array}{r} 203 \\ +302 \\ \hline \end{array}$$

8.
$$\begin{array}{r} 713 \\ +272 \\ \hline \end{array}$$

Solve for the unknown.

9. $6 + X = 8$ 10. $4 + R = 9$

Fill in the blanks with equal addends to make the double. The sum is an even number.

11. _____ + _____ = 14

12. Isabella has 40 pennies. Her brother wants to trade 8 nickels for her pennies. Should she say yes?

   _____

13. Five children came in out of the rain. Skip count to find how many feet were tracking mud into the house.

   _____

14. Rodney has 15 comic books and Eugene has 16. How many do they have altogether?

   _____

15. The bus driver picked up six passengers at his first stop and two at his second stop. Then he picked up six more passengers at the third stop. How many passengers has he picked up in all?

   _____

Skip count by five and two. Write the numbers.

1. ____, 10, ____, ____, ____, ____, ____, ____, 45, ____

2. ____, ____, ____, 8, ____, ____, 14, ____, ____, ____

Add. Regroup if needed.

3.
```
    2 7
  + 3 3
```

4.
```
    8 1
  +   3
```

5.
```
    3 6
  + 1 4
```

Add. These do not need regrouping.

6.
```
    2 9 3
  + 1 0 4
```

7.
```
    6 4 5
  + 3 2 1
```

8.
```
    7 8 4
  + 2 1 5
```

Solve for the unknown.

9. $9 + B = 10$

10. $3 + Y = 9$

11. $7 + G = 12$

12. Vontoria needs 45¢ to buy a candy bar. How many nickels does she need?

_____

13. Playing catch with his dad, Raleigh caught the ball 25 times before lunch and 16 times after lunch. How many times did he catch the ball in all?

_____

14. Shane's favorite baseball team had 4 runs in the first inning, 2 runs in the second inning, and 8 in the third inning. How many runs did the team have during the first three innings?

_____

15. Which is worth more: five nickels or four dimes?

_____

Skip count by five and ten. Write the numbers.

1. ____, ____, ____, 20, ____, ____, ____, ____, ____, ____

2. ____, 20, ____, ____, ____, ____, ____, ____, ____, ____

Add. Regroup if needed.

3.
```
  5 7
+ 2 2
```

4.
```
  7 4
+   6
```

5.
```
  2 4
+ 1 8
```

Add. These do not need regrouping.

6.
```
  6 8 0
+ 1 1 9
```

7.
```
  5 3 2
+ 2 2 2
```

8.
```
  1 9 2
+ 2 0 7
```

Solve for the unknown.

9. 8 + R = 10   10. 7 + A = 15

11. 9 + U = 12

Draw lines to match the questions with the right answers.

12. Which coin is worth ten cents?

13. Which coin is worth one cent?

14. Which coin is worth five cents?

15. Mom bought a chair for 314 dollars and a table for 322 dollars. How much did Mom spend on furniture?

_____

APPLICATION AND ENRICHMENT

Skip count by 5 to 100.

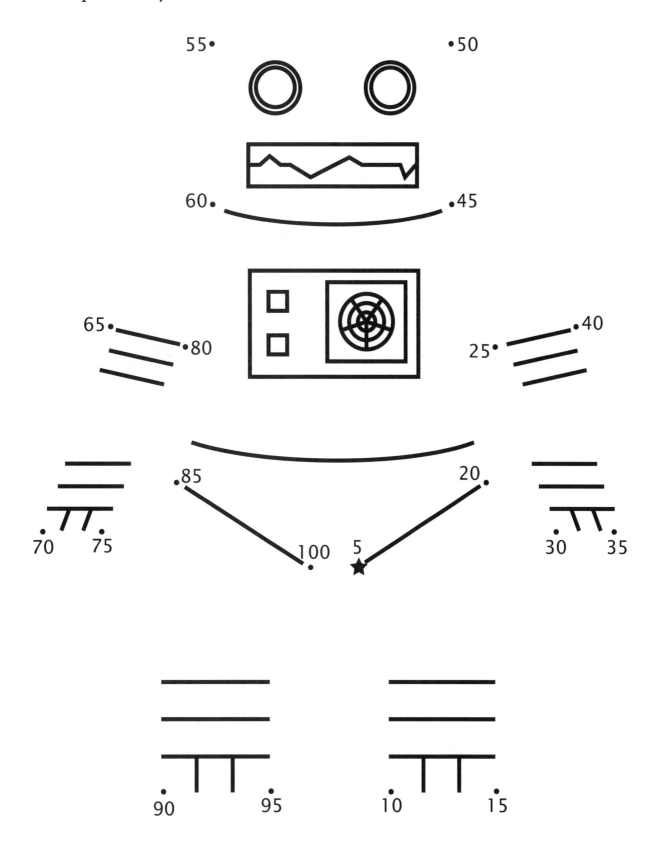

Match each thing with the money you need to buy it.

15¢

30¢

25¢

5¢

Fill in the blanks and say the amount.

1.

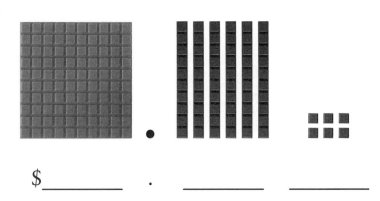

$\$_____$. $_____$ $_____$

2.

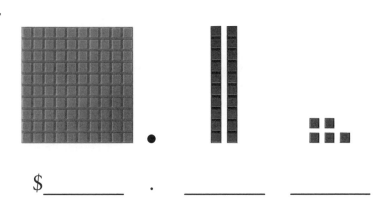

$\$_____$. $_____$ $_____$

3.

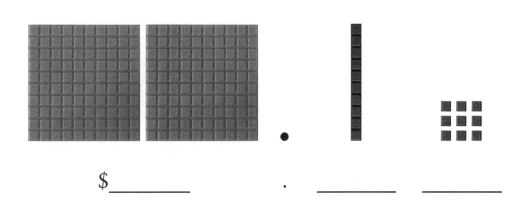

$\$_____$. $_____$ $_____$

4.

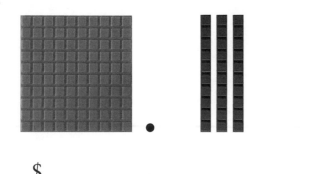

$_____   .   _____   _____

Build and say.

5. $2.62

6. $2.05

7. $1.96

8. $3.18

Fill in the blanks and say the amount.

1.

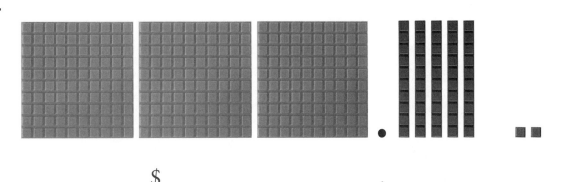

$\underline{}$. $\underline{}$ $\underline{}$

2.

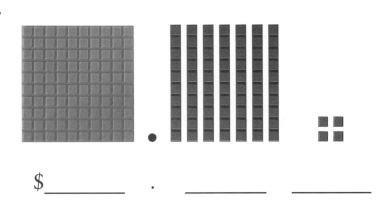

$\underline{}$. $\underline{}$ $\underline{}$

3.

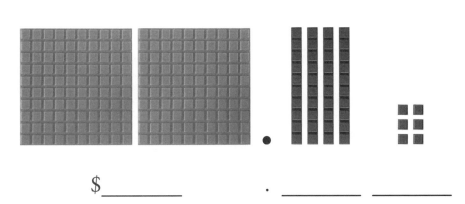

$\underline{}$. $\underline{}$ $\underline{}$

Build and say.

4. $4.51

5. $3.60

6. $1.81

7. $2.07

8. Erica has two dollars, one dime, and five pennies. How much money does she have?

_____

Fill in the blanks and say the amount.

1.

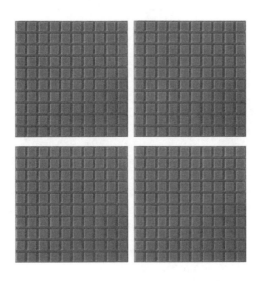

$_____   •   _____ ■

$_____   . _____ _____

2.

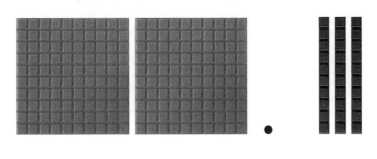

$_____   . _____ _____

3.

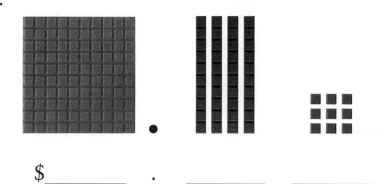

$_____ . _____ _____

Build and say.

4. $3.23

5. $1.08

6. $4.52

7. $2.90

8. Rhonda found six dollars, seven dimes, and three pennies. How much money did she find?

_____

SYSTEMATIC REVIEW

Fill in the blanks and say the amount.

1.

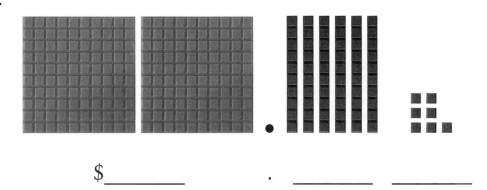

$ _____ . _____ _____

Build and say.

2. $1.48   3. $2.73

4. $4.05   5. $3.60

Skip count to find how many cents.

6.   = _____ ¢

Add.

7.
```
   4 9
 +   9
 ─────
```

8.
```
   3 1 1
 + 2 3 8
 ───────
```

9. $\begin{array}{r} 6\,5 \\ +\,2\,5 \\ \hline \end{array}$

Solve for the unknown.

10. $7 + S = 13$

11. $3 + B = 5$

12. $4 + V = 12$

13. Amanda's dad gave her five dollars and four pennies. How much money did she receive?

    _____

14. When Dan went to the amusement park, he rode the big roller coaster 19 times and the kiddie roller coaster 25 times. How many roller coaster rides did Dan have?

    _____

15. Elizabeth bought four birthday cards and four get-well cards. The next day she bought nine Christmas cards. How many cards did she buy in all?

    _____

Fill in the blanks and say the amount.

1.

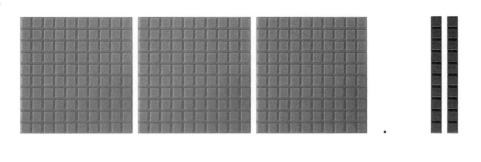

$\$$ _____   .   _____ _____

Build and say.

2. $2.31   3. $4.55

4. $1.06   5. $3.78

Skip count to find how many cents.

6.

= ____¢

Add.

7.   17
    + 18

8.
$$\begin{array}{r} 555 \\ +\ 132 \\ \hline \end{array}$$

9.
$$\begin{array}{r} 49 \\ +\ 34 \\ \hline \end{array}$$

Round to the nearest tens place.

10. $76 \rightarrow$ _____

11. $13 \rightarrow$ _____

12. $25 \rightarrow$ _____

13. While swinging on the playground, Andrew lost nine nickels from his pocket. Use a decimal point when writing how many cents he lost.

_____

14. On Mother's Day, Aiden bought his mother five white roses and two yellow roses. Then he decided to add two red roses to the bunch. How many roses did he buy in all?

_____

15. Al has three dimes. How many cents does he have?

_____

Fill in the blanks and say the amount.

1.

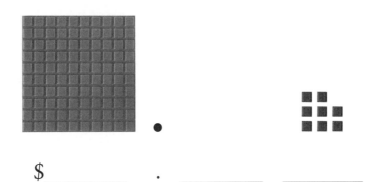

$\$$_____ . _____ _____

Build and say.

2. $1.16

3. $3.09

4. $2.65

5. $4.70

Skip count to find how many cents.

6. = ____¢

Add.

7.   9 2
   + 4

8.   3 3 7
   + 2 0 2

9.
```
   6 1
 + 2 9
```

Compare. Then fill in the oval with <, >, or =.

10. 3 + 7 $\bigcirc$ 6 + 1

11. 4 + 4 $\bigcirc$ 8

12. 27 $\bigcirc$ 72

13. Douglas earned eight dollars, six dimes, and nine pennies. How much money did he earn?

_____

14. Jonathan walked three miles and ran four miles. The next day he rode his bicycle eight miles. How many miles did Jonathan go altogether?

_____

15. Lindsay sailed with all her sails up for 29 miles. When the wind picked up, she took in one of the sails and sailed for another 18 miles. How many miles did she sail altogether?

_____

10G

Use a blue crayon to color all the boxes that have numbers that you say when you count by fives. See if you can go all the way to 100.

Next, use your red crayon to color all the boxes that have numbers that you say when you count by tens. What color are the tens boxes now?

0	1	2	3	4	5	6	7	8	9
10	11	12	13	14	15	16	17	18	19
20	21	22	23	24	25	26	27	28	29
30	31	32	33	34	35	36	37	38	39
40	41	42	43	44	45	46	47	48	49
50	51	52	53	54	55	56	57	58	59
60	61	62	63	64	65	66	67	68	69
70	71	72	73	74	75	76	77	78	79
80	81	82	83	84	85	86	87	88	89
90	91	92	93	94	95	96	97	98	99
100									

Match each thing with the money you need to buy it.

$2.62

$1.24

$3.05

$2.26

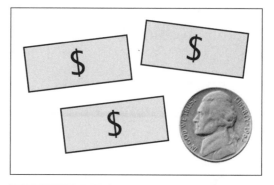

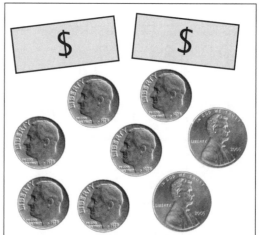

11A

Round to the nearest hundred.

1. 190 → _____

2. 206 → _____

3. 355 → _____

Round to the nearest hundred and estimate the answer. Then find the exact answer.* The first one has been done for you.

4.
```
  1 1
  3 6 4    (400)
+ 2 9 7   +(300)
  6 6 1    (700)
```

5.
```
  6 2 8    (    )
+ 1 7 5   +(    )
           (    )
```

6.
```
  3 5 9    (    )
+ 2 5 4   +(    )
           (    )
```

7.
```
  5 3 7    (    )
+ 2 3 3   +(    )
           (    )
```

*The estimate may be close to 100 more or less than the exact answer.

8.　　1 6 8　　(　　)
　　+ 4 5 2　　+ (　　)
　　　　　　　　(　　)

9.　　1 2 3　　(　　)
　　+　 8 8　　+ (　　)
　　　　　　　　(　　)

10.　　6 7 6　　(　　)
　　 + 1 4 5　　+ (　　)
　　　　　　　　(　　)

11.　　2 9 9　　(　　)
　　 + 3 1 1　　+ (　　)
　　　　　　　　(　　)

12. First, 124 lights burned out on the big Christmas tree downtown. Then 176 more lights burned out. How many lights need to be replaced?

_____

Round to the nearest hundred.

1. 476 → _____

2. 515 → _____

3. 610 → _____

Round to the nearest hundred and estimate the answer. Then find the exact answer.

4.   359   (   )
    + 1 2 6   + (   )
              (   )

5.   1 3 8   (   )
    + 2 1 2   + (   )
              (   )

6.   1 5 7   (   )
    + 1 4 2   + (   )
              (   )

*7.   2 2 7   (   )
     +   3 9   + (   )
               (   )

*Round 39 to 0, as it is closer to 0 than to 100.

8.　　4 4 9　　(　　)
　　　+ 1 3 7　　+ (　　)
　　　　　　　　　(　　)

9.　　2 3 5　　(　　)
　　　+ 1 4 5　　+ (　　)
　　　　　　　　　(　　)

10.　　1 0 9　　(　　)
　　　+ 2 0 7　　+ (　　)
　　　　　　　　　(　　)

11.　　4 1 6　　(　　)
　　　+ 3 2 9　　+ (　　)
　　　　　　　　　(　　)

12. On Monday Steve read 123 pages of his book. On Tuesday he read 169 pages. How many pages has he read in all?

_____

Round to the nearest hundred.

1. 450 → _____

2. 103 → _____

3. 278 → _____

Round to the nearest hundred and estimate the answer. Then find the exact answer.

4.   217   (   )
  + 324 + (   )
          (   )

5. 266   (   )
  + 18 + (   )
        (   )

6.   134   (   )
  + 365 + (   )
          (   )

7.   119   (   )
  + 207 + (   )
          (   )

8.   555   (   )
    + 3 4 8   + (   )
             (   )

9.   806   (   )
    + 1 0 6   + (   )
             (   )

10.   119   (   )
    + 2 1 7   + (   )
             (   )

11.   248   (   )
    + 2 5 2   + (   )
             (   )

12. On our trip to Grandma's house, Dad drove 263 miles, and Mom drove 179 miles. How far is it to Grandma's house?

_____

11D

Round to the nearest hundred.

1. $755 \rightarrow$ _____

2. $115 \rightarrow$ _____

Add. Regroup if needed.

3.
```
  806
+ 106
```

4.
```
  248
+ 252
```

5.
```
  337
+ 172
```

6.
```
  54
+ 28
```

7.
```
  53
+ 37
```

8.
```
  18
+ 29
```

These subtraction facts review subtracting 0, 1, and 2.

9.
```
  1
- 1
```

10.
```
  10
-  2
```

11.
$$\begin{array}{r} 8 \\ -\ 1 \\ \hline \end{array}$$

12.
$$\begin{array}{r} 3 \\ -\ 0 \\ \hline \end{array}$$

13.
$$\begin{array}{r} 4 \\ -\ 3 \\ \hline \end{array}$$

14.
$$\begin{array}{r} 6 \\ -\ 2 \\ \hline \end{array}$$

15.
$$\begin{array}{r} 5 \\ -\ 4 \\ \hline \end{array}$$

16.
$$\begin{array}{r} 8 \\ -\ 2 \\ \hline \end{array}$$

Skip count by two and write the numbers.

17. ____, 4, ____, ____, 10, ____, ____, ____, ____, ____

18. Andrew has five dollars and twenty-six cents. Use a decimal point and dollar sign to write that amount.

   _____

19. We traveled 55 miles last week and 78 miles this week. How many miles did we travel those two weeks?

   _____

20. Jim had $145 in his savings. He got $56 for his birthday. How much money does Jim have now?

   _____

Round to the nearest hundred.

1. 361 → ____

2. 209 → ____

Add. Regroup if needed.

3.
```
  2 3 5
+ 3 6 5
```

4.
```
  3 0 0
+ 4 0 9
```

5.
```
  2 4 9
+ 1 3 2
```

6.
```
  2 8
+ 3 8
```

7.
```
  6 5
+ 3 5
```

8.
```
  5 8
+ 4 2
```

These subtraction facts review subtracting 0, 1, and 2.

9.
```
  4
- 2
```

10.
```
  7
- 2
```

11. $\begin{array}{r} 3 \\ -\ 1 \\ \hline \end{array}$ 12. $\begin{array}{r} 11 \\ -\ 2 \\ \hline \end{array}$

13. $\begin{array}{r} 6 \\ -\ 5 \\ \hline \end{array}$ 14. $\begin{array}{r} 8 \\ -\ 0 \\ \hline \end{array}$

15. $\begin{array}{r} 10 \\ -\ 9 \\ \hline \end{array}$ 16. $\begin{array}{r} 9 \\ -\ 2 \\ \hline \end{array}$

Skip count by five and write the numbers.

17. ____, ____, 15, ____, ____, ____, ____, ____, ____, ____

18. A tree grew 17 inches one year and 9 inches the next year. How much did it grow during those two years?

   _____

19. Captain Cook spotted 138 penguins in the water and 256 on the shore. How many penguins did he see?

   _____

20. Mom had five eggs. Two of the eggs were cracked. How many eggs were not cracked? (Watch for word problems that review subtraction.)

   _____

Round to the nearest hundred.

1. 519 → _____

2. 682 → _____

Add. Regroup if needed.

3.
```
   4 2 9
 + 2 6 6
```

4.
```
   1 0 1
 +   8 9
```

5.
```
   2 3 8
 + 2 4 3
```

6.
```
   9 2
 +  8
```

7.
```
   4 8
 + 3 2
```

8.
```
   6 3
 + 2 7
```

These facts review subtracting two and difference of two.

9.
```
   5
 − 3
```

10.
```
   1 0
 −  2
```

11.   7
     − 5

12.   6
     − 2

13.   9
     − 7

14.   8
     − 2

15.   1 0
     − 8

16.   1 1
     − 9

Skip count by ten and write the numbers.

17. _____, _____, _____, 40, _____, _____, _____, _____, _____, _____

18. Deb has 21 jelly beans and 48 mints. Round to the nearest 10 and estimate how many pieces of candy she has.

_____

19. Pete drove 512 miles one day and 345 the next day. How far did he drive? Estimate first and then solve.

_____

20. Drew lost eight pennies. If he finds six pennies, how many are still lost? Use a decimal point and a dollar sign to write the answer.

_____

11G

Here are some challenging addition squares. Add across and down.

35	21	
15	42	

16	52	
28	17	

64	5	
6	55	

47	11	
12	61	

25	70	
8	13	

20	30	
40	50	

Color by number: adding whole hundreds

100, 200 = yellow
300, 400 = brown
500, 600, 700 = green
800, 900, 1000 = blue

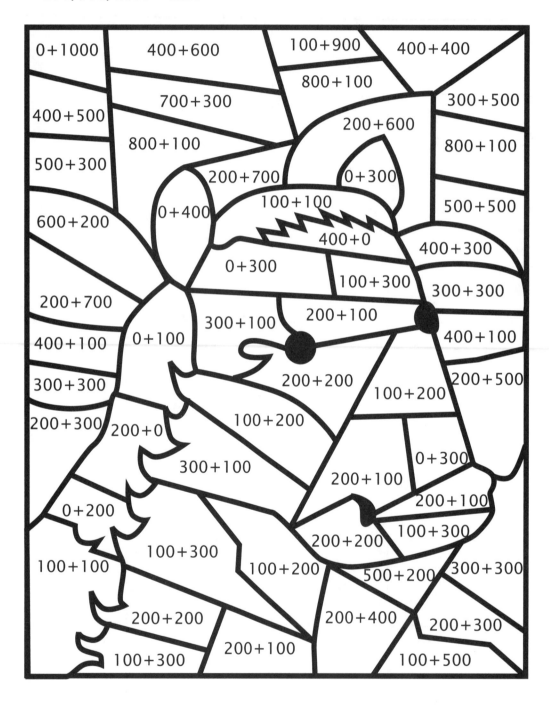

Add the money. The first one has been done for you.

1.
```
     1 1
   $3.2 1
 +  4.9 9
   $8.2 0
```

2.
```
   $7.0 9
 +  1.9 2
```

3.
```
   $3.3 3
 +  1.4 4
```

4.
```
   $6.5 0
 +  2.7 7
```

5.
```
   $4.0 0
 +  2.5 1
```

6.
```
   $5.1 9
 +  1.3 8
```

7.
```
   $1.0 0
 +  0.7 5
```

8.
```
   $2.0 3
 +  1.9 0
```

9. $8.7 5
    + 0.8 0
    ⎯⎯⎯⎯

10. Rose went shopping. If she spent $5.25 in one store and $3.38 in another store, how much did she spend in all?

    ⎯⎯⎯⎯⎯⎯⎯⎯⎯⎯⎯⎯⎯

11. Chance had $2.63 in his pocket. He got $5.50 in a birthday card. How much money does he have now?

    ⎯⎯⎯⎯⎯⎯⎯⎯⎯⎯⎯⎯⎯

12. Joseph bought toys for his pug dog, Max. If he spent $2.99 for a rubber bone and $3.61 for a ball, how much did he spend in all?

    ⎯⎯⎯⎯⎯⎯⎯⎯⎯⎯⎯⎯⎯

Add the money.

1. $7.6 5
   + 0.6 0

2. $6.3 1
   + 1.2 9

3. $5.8 3
   + 0.2 4

4. $3.1 9
   + 0.9 0

5. $2.0 0
   + 0.9 8

6. $1.0 3
   + 1.2 5

7. $3.7 2
   + 4.0 8

8. $1.9 9
   + 1.8 2

9. $2.8 7
    + 6.8 9
    _____

10. Elizabeth found $3.45 in her drawer and $1.99 behind her dresser. How much money did she find in all?

    _____

11. Meredith wants to buy a book that costs $5.55 and a box of crayons that costs $2.15. How much money does she need?

    _____

12. Daniel's brother gave him $6.34. His sister gave him $2.95. How much money was Daniel given altogether?

    _____

LESSON PRACTICE

Add the money.

1. $2.1 3
   + 1.9 2

2. $4.7 1
   + 1.3 6

3. $6.4 1
   + 0.3 9

4. $5.0 0
   + 2.5 0

5. $6.6 3
   + 2.4 4

6. $7.3 5
   + 1.0 5

7. $1.6 3
   + 0.7 2

8. $4.9 9
   + 3.7 9

9. $6.3 3
     + 2.9 1
    _____

10. Caleb lost $5.10 yesterday and $3.91 today. How much did he lose altogether?

    _____

11. Caleb found $2.50 of his lost money (see #10). Timothy felt sorry for him and gave him $4.50. How much money does Caleb have now?

    _____

12. Thomas has $3.72 in his left pocket and $3.68 in his right pocket. How much money does he have? Does he have enough to buy a toy that costs $6.00?

    _____

Add. Regroup if needed.

1. $1.66
    + 4.08

2. $3.09
    + 2.56

3. $3.57
    + 2.62

4.   422
    + 389

5.   19
    + 16

6.   17
    + 25

These facts review subtracting 9.

7.   12
    - 9

8.   18
    - 9

9.   9
    - 9

10. 14
    - 9

11. 17
    - 9

12. 13
    - 9

13. 16
     − 9

14. 15
     − 9

Fill in the blanks and say the amount.

15.

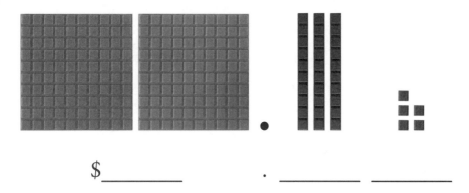

$_____ . _____ _____

16. Nicholas wrapped 9 Christmas gifts. Two gifts are left to wrap. How many gifts did he have to start with? Solve for G.
G − 9 = 2 gifts

_____

17. Forty-six grown-ups and sixty-three children are coming to the wedding. Estimate how many chairs are needed. Then add to find the exact number.

_____

18. Jacob spent $2.78 for a sandwich and $1.19 for a drink. How much did he spend in all?

_____

Add. Regroup if needed.

1. $1.68
   + 4.77

2. $4.56
   + 4.44

3. $2.63
   + 0.51

4. 684
   + 122

5. 62
   + 29

6. 83
   + 7

These facts review subtracting 8.

7. 11
   - 8

8. 9
   - 8

9. 17
   - 8

10. 12
    - 8

11. 14
    - 8

12. 13
    - 8

13. 1 5
    − 8
    ‾‾‾

14. 1 6
    − 8
    ‾‾‾

Fill in the blanks and say the amount.

15.

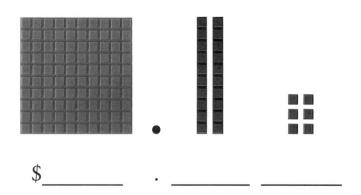

$_____ . _____ _____

16. Emily lost 8 dimes. If she has 5 dimes left, how many did she have to start with? Solve for the unknown. Set this up as you did with #16 on 12D.

_____

17. Ava wrapped 267 candies with nuts and 197 plain candies to sell in her dad's shop. How many candies did she wrap in all? Round to the nearest hundred and estimate. Then solve the problem.

_____

18. Michael has 3 dimes and 5 nickels. Find out how much money he has in all. Use a dollar sign and decimal point to write your answer.

_____

Add. Regroup if needed.

1. $2.78
   + 6.58

2. $3.52
   + 1.77

3. $8.91
   + 0.05

4. 379
   + 264

5. 54
   + 18

6. 47
   + 9

These facts review subtracting 8 and doubles.

7. 12
   − 6

8. 13
   − 8

9. 10
   − 5

10. 14
   − 7

11. 12
   − 8

12. 17
   − 8

13.
$$\begin{array}{r} 8 \\ -\ 4 \\ \hline \end{array}$$

14.
$$\begin{array}{r} 6 \\ -\ 3 \\ \hline \end{array}$$

Fill in the blanks and say the amount.

15.

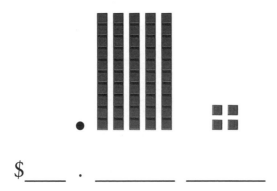

$\_\_\_\_$. $_____$ $_____$

16. I watched 8 birds fly away from my bird feeder. If 8 birds are left on the feeder, how many were there to start with? Solve for the unknown.

_____

17. Bria has three nickels and five pennies. How much money does she have in all? Use a dollar sign and decimal point to write your answer.

_____

18. Kara talked on the phone 35 minutes yesterday and 17 minutes today. How many minutes did she talk in all?

_____

Answer the questions and fill in the crossword puzzle.

Across

2. The coin worth one cent is a _____ .

4. A _____ is worth one hundred cents.

7. The coin worth ten cents is a ____ .

Down

1. How many cents are in a dime? _____

3. The coin worth five cents is a _____ .

5. How many cents are in a penny? _____

6. How many cents are in a nickel? _____

Draw pictures to help you review subtraction facts. Cross out the pictures you subtract each time.

1. Jim blew 18 soap bubbles. Draw the bubbles.

2. Jeff popped nine bubbles. How many bubbles are left? _____

3. The cat broke three bubbles. How many bubbles are left? _____

4. Four bubbles landed on the table and broke. How many bubbles are left? _____

5. Two bubbles floated out the door and disappeared. Now how many bubbles are left? _____

Add the columns. Find 10 first if you can.

1.   2
      8
    + 5

2.   6
      3
    + 4

3.   5
      4
    + 5

4.   8
      2
      6
    + 1

5.   6
      2
      4
    + 7

6.   3
      7
      5
    + 5

Add the columns. Find 10 first if you can.

7.　　4 0
　　　　5 0
　　　+ 1 0

8.　　3 0
　　　　7 0
　　　+ 4 0

9.　　5 1
　　　　5 9
　　　+ 2 0

Add. First look for 10.

10.　9 + 1 + 8 + 2 = _____

11.　6 + 1 + 8 + 4 + 2 = _____

12.　Bill got 4 presents for his birthday from his grandparents, 5 from his parents, 6 from his brothers and sisters, and 5 from his friends. How many presents did he get in all?

　　　_____

Add the columns. Find 10 first if you can.

1.   6
      4
   + 9

2.   9
      5
   + 5

3.   7
      2
   + 3

4.   8
      5
      4
   + 2

5.   9
      1
      7
   + 3

6.   1
      2
      3
   + 9

7.
```
  6 0
  4 0
+ 2 0
```

8.
```
  2 0
  6 0
+ 8 0
```

9.
```
  4 3
  2 7
+ 6 1
```

Add. First look for 10.

10. 8 + 2 + 3 + 7 + 2 = _____

11. 4 + 9 + 6 + 2 + 1 = _____

12. For gym class we ran laps around the field. Jason ran 8 laps, Mary ran 2 laps, Susan ran 3 laps, and Eric ran 3 laps. How many laps were run in all?

_____

LESSON PRACTICE

Add the columns. Find 10 first if you can.

1.
$$\begin{array}{r} 2 \\ 8 \\ + 7 \\ \hline \end{array}$$

2.
$$\begin{array}{r} 5 \\ 1 \\ + 5 \\ \hline \end{array}$$

3.
$$\begin{array}{r} 4 \\ 3 \\ + 3 \\ \hline \end{array}$$

4.
$$\begin{array}{r} 2 \\ 8 \\ 2 \\ + 3 \\ \hline \end{array}$$

5.
$$\begin{array}{r} 3 \\ 8 \\ 7 \\ + 2 \\ \hline \end{array}$$

6.
$$\begin{array}{r} 9 \\ 5 \\ 6 \\ + 1 \\ \hline \end{array}$$

7.
```
    8 0
    5 0
    1 0
  + 1 0
  ─────
```

8.
```
    4 0
    6 0
      4
  +   2
  ─────
```

9.
```
    8 4
    2 6
    1 7
  + 2 3
  ─────
```

Add. First look for 10.

10. $6 + 7 + 3 + 4 + 6 =$ _____

11. $5 + 5 + 4 + 3 + 7 =$ _____

12. Mrs. Green's class had a reading contest. Here are the numbers of books read by the different students: 11, 14, 5, 10, and 4. How many books were read altogether?

_____

Add the columns. Find 10 first if you can.

1.
```
    3
    7
  + 6
  ___
```

2.
```
    3
    9
    1
  + 7
  ___
```

3.
```
    4 2
    6 4
    8 2
  + 2
  _____
```

4.
```
   $2.8 5
  + 6.5 6
  _____
```

5.
```
    1 4 9
  + 2 7 3
  _____
```

6.
```
    1 4
  + 9
  _____
```

These facts review subtracting 9 and 10.

7.
```
   1 0
  - 8
  _____
```

8.
```
   1 0
  - 4
  _____
```

9.
```
    9
  - 6
  _____
```

10.
```
   1 0
  - 7
  _____
```

11. 9
    − 2

12. 9
    − 4

13. 9
    − 1

14. 1 0
    − 6

Skip count and write the numbers.

15. ____, 20, ____, 40, ____, ____, ____, ____, ____, ____

QUICK REVIEW

A triangle has three sides. A rectangle has four sides and four square corners. A rectangle with all four sides the same length is called a square.

triangle   square   rectangle

16. Circle the name of the shape with three sides.

    square   rectangle   triangle

17. A doctor helped 13 patients on Wednesday, 15 on Thursday, and 11 on Friday. How many patients did he help in all?

    _____

18. Ruth has $6.18, and her mother gave her $2.00. Does she have enough money to buy a toy that costs $9.00?

    _____

Add the columns. Find 10 first if you can.

1.
```
   2
   4
 + 2
```

2.
```
   6
   3
   8
 + 2
```

3.
```
  1 0
  1 0
  2 0
 +2 0
```

4.
```
  $3.4 6
 + 2.5 4
```

5.
```
  1 0 0
 +2 7 8
```

6.
```
  7 9
 +8 8
```

These subtraction facts review subtracting 9 and 10.

7.
```
  1 0
 - 2
```

8.
```
  1 0
 - 5
```

9.
```
   9
 - 5
```

10.
```
   9
 - 7
```

11. 1 0
     − 3
     _____

12. 9
     − 8
     _____

13. 9
     − 3
     _____

14. 1 0
     − 4
     _____

Skip count and write the numbers.

15. _____, _____, 6, _____, 10, _____, _____, _____, _____, _____

16. Circle the names of the shapes with four sides.

     square   triangle   rectangle

17. Julia had an ice cream shop. She sold 35 small cones, 22
     large cones, and 12 extra large cones. How many ice cream
     cones did she sell in all?

     _____

18. David has nine nickels and seven pennies. How much money
     does he have?

     _____

Add the columns. Find 10 first if you can.

1.
```
   5
   5
 + 2
```

2.
```
   5
   6
   4
 + 7
```

3.
```
   2 3
   3 7
   2 3
 + 1 2
```

4.
```
  $4.3 6
 + 1.2 2
```

5.
```
   1 1 6
 +   6 8
```

6.
```
   1 6
 + 4 4
```

Review these subtraction facts.

7.
```
   1 7
 -   8
```

8.
```
   1 4
 -   7
```

9.
```
   1 0
 -   3
```

10.
```
    9
 -  5
```

11.　1 5
　　　－ 9

12.　　9
　　　－ 7

13.　1 2
　　　－ 8

14.　1 6
　　　－ 9

Skip count and write the numbers.

15. ____, ____, 15, ____, ____, ____, ____, ____, ____, 50

Match the shapes with their names as given on 13D.

16.　　　　　　　　rectangle

17.　　　　　　　　triangle

18.　　　　　　　　square

19. Dan has six nickels, five dimes, and four pennies. How much money does he have altogether?

_____

20. Timothy counted 33 raisins in his cereal bowl, and Peter counted 49 raisins in his bowl. How many raisins did the boys count altogether?

_____

13G

Match each thing with the money you need to buy it.

$1.31

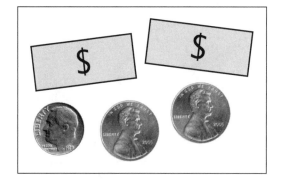

$2.12

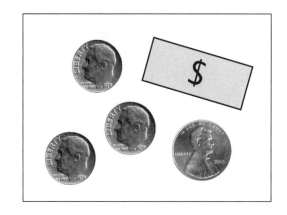

45¢

$2.15

How much money do you need to buy both of these things? Circle the answer.

$1.25

$1.29

How much money do you need to buy both of these things?
Circle the answer.

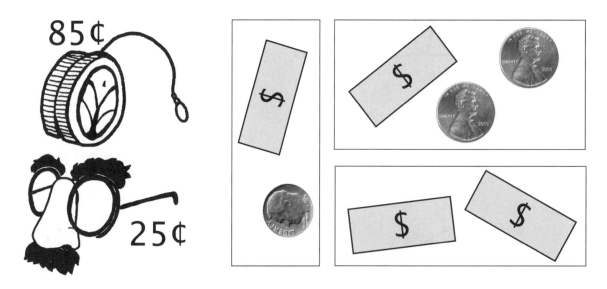

85¢

25¢

The two block is one inch long. Use it to measure #1 and #2.

1.

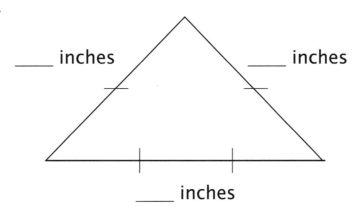

____ inches   ____ inches

____ inches

2.

____ inches

Measure using a ruler.

3.

_____ "

4.

_____ "

Is line 3 or line 4 longer? ____

How much longer is it? ____ − ____ = ____ inches longer

5.

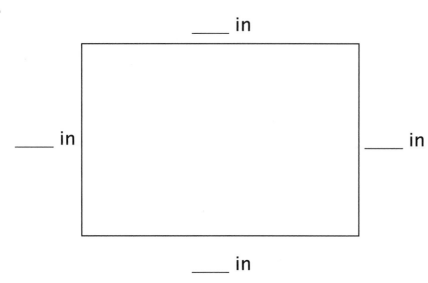

_____ in

_____ in

_____ in

_____ in

6. 12 inches = _____ foot

The two block is one inch long. Use it to measure #1 and #2.

1.

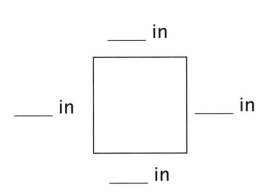

____ in

____ in   ____ in

____ in

2.

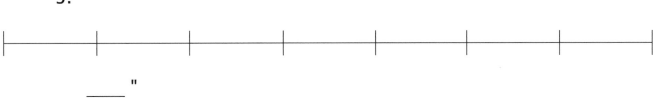

____ inches

Measure using a ruler.

3.

____ "

4.

____ "

Is line 3 or line 4 longer?   ____

How much longer is it?   ____ – ____ = ____ inch longer

5.

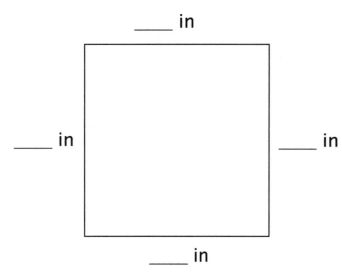

6. ____ inches = 1 foot

The two block is one inch long. Use it to measure #1 and #2.

1.

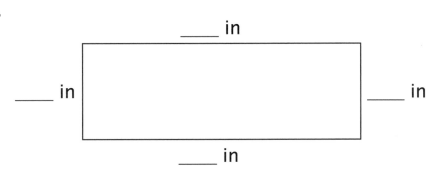

____ in

____ in

____ in

____ in

2.

____ inches

Measure using a ruler.

3.

____ "

4.

____ "

Is line 3 or line 4 longer? ____

How much longer is it? ____ – ____ = ____ inches longer

5.

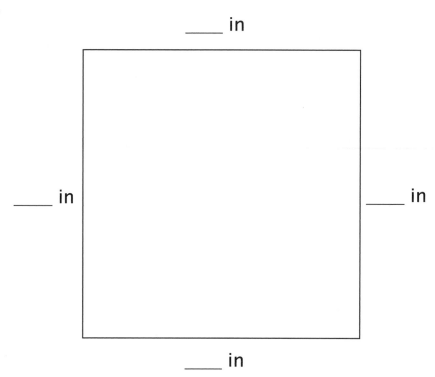

_____ in

6. 2 feet = _____ inches (Add 12 + 12 to find the answer.)

Measure using a ruler.

1.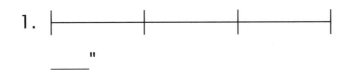

    ____ "

2.

    ____ "

Is line 1 or line 2 longer? ____

How much longer is it? ____ – ____ = ____ inches longer

Add.

3. $1.7 7
    + 2.7 8

4.   1 1 8
    + 1 2 2

5.   2 3
    + 2 4

6. $1 + 2 + 3 + 7 + 8 + 9 =$ _____

7. $6 + 6 + 4 + 3 + 7 =$ _____

Compare. Then fill in the oval with <, >, or =.

8. $10 - 4$ ◯ $3 + 3$ 9. $8 - 2$ ◯ $5 + 3$

Review these subtraction facts. Subtraction may be written horizontally.

10. $7 - 3 =$ _____   11. $8 - 5 =$ _____

12. $7 - 4 =$ _____   13. $8 - 3 =$ _____

14. $9 - 5 =$ _____   15. $15 - 9 =$ _____

16. How many sides does a triangle have?

_____

17. Jared has six dimes. If he lost three dimes, how many cents does he have left?

_____

18. Mom bought 6 apples, 5 oranges, 10 bananas, and 4 pears. How many pieces of fruit did she buy?

_____

Measure using a ruler.

1.

_____ "

2. |——————|——————|

_____ "

Is line 1 or line 2 longer? _____

How much longer is it? _____ – _____ = _____ inches longer

Add.

3. $3.6 3
    + 2.7 7

4.   4 5 2
    + 1 8 1

5.   8 9
    + 8 7

6. 3 + 3 + 7 + 3 + 2 = _____

7. 10 + 2 + 10 + 2 = _____

Compare. Then fill in the oval with <, >, or =.

8. 7 - 3 ◯ 7 - 4

Subtraction review

9. 11 - 7 = ____   10. 13 - 7 = ____

11. 8 - 3 = ____   12. 16 - 7 = ____

13. 12 - 7 = ____   14. 15 - 7 = ____

15. How many sides does a rectangle have?

   _____

16. Bonnie flew 256 miles in her airplane. After she landed to get fuel, she took off again and flew another 289 miles. How far did she fly in all?

   _____

17. Peter called Debbie three times, his mother four times, the doctor one time, and the library two times. How many calls did Peter make in all?

   _____

18. Sam is three feet tall. How many inches tall is he? (Use column addition.)

   _____

Measure using a ruler.

1.

___"

2.

___"

Is line 1 or line 2 longer? ____

How much longer is it? ____ – ____ = ____ inch longer

Add.

3. $1.5 6
    + 2.3 8

4.   4 1 9
    + 4 1 9

5.   3 9
    + 4 2

6. 5 + 5 + 5 + 5 = ____

7. 3 + 5 + 10 = ____

Compare. Then fill in the oval with <, >, or =.

8. 14 − 9 $\bigcirc$ 4 + 4

Review these subtraction facts.

9. 11 − 6 = ____   10. 14 − 6 = ____

11. 15 − 6 = ____   12. 13 − 6 = ____

13. 15 − 7 = ____   14. 7 − 4 = ____

15. How many sides does a square have?

    _____

16. Matthew has seven dimes, and his sister has three nickels. How much money do they have together?

    _____

17. Bill's book has 13 pages. He read three pages before lunch and three pages after lunch. How many pages does he have left to read?

    _____

18. Jerry traveled 56 miles the first day and 35 miles the second day. How many miles did Jerry travel in those two days?

    _____

BETA

Each of these measurement activities has three parts. First, decide if you are going to give your answer with inches or with feet. Next, estimate the answer—give your best guess! Finally, measure and write the answer. Don't forget to write feet or inches after the answer.

1. How long is your thumb?

    Is feet or inches the best unit of length? _____

    Estimate the answer. _____

    Measure your thumb and give your final answer. _____

2. How long is the room you are in now?

    Is feet or inches the best unit of length? _____

    Estimate the answer. _____

    Measure the room and give your final answer. _____

3. Choose a favorite book. How wide is the book?

    Is feet or inches the best unit of length? _____

    Estimate the answer. _____

    Measure the book and give your final answer. _____

Note: Encourage the child to choose the nearest whole inch or foot for the answer. You can point out that this is like rounding numbers. We will teach measurement with fractions of an inch in *Epsilon*.

This enrichment will help you practice measuring and comparing. Write the first measurement in the first box for each problem. Write the second measurement in the last box. Remember to include measurement units.

Compare the measures. Write >, <, or = in the oval.

1. How long is your foot?

   How long is your mom or dad's foot?

2. How tall is your math book?

   How wide is your math book?

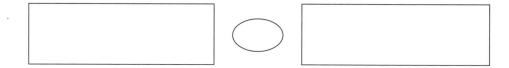

3. How long is the fork you eat with?

   How long is the spoon you eat with?

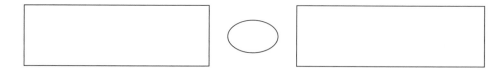

Circle the name of each shape. Count the spaces on each side and write the length in the box. Find the peRIMeter of each shape. (Notice that we have drawn the inches smaller than they really are.) Use "square" for the shape with four sides that are the same length.

1. The shape is a:

square   (rectangle)   triangle

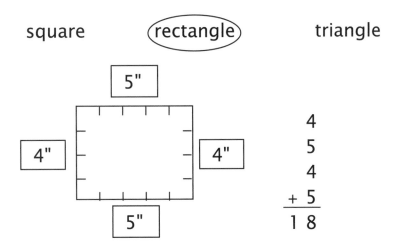

```
    4
    5
    4
  + 5
  ─────
  1 8
```

The perimeter is __18__ inches.

2. The shape is a:

square   circle   triangle

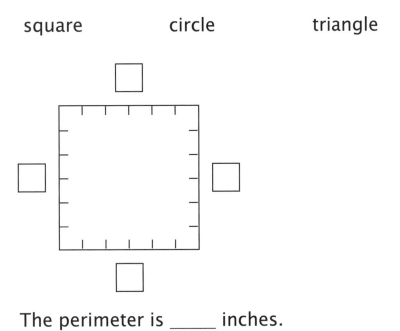

The perimeter is _____ inches.

3. The shape is a:

square   rectangle   triangle

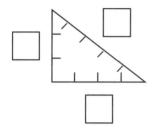

The perimeter is _____ inches.

4. The shape is a:

square   rectangle   triangle

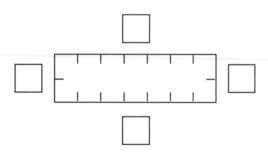

The perimeter is _____ inches.

Circle the name of each shape. Count the spaces on each side and write the length in the box. Add up the peRIMeter (distance around) each shape.

1. The shape is a:

square   rectangle   triangle

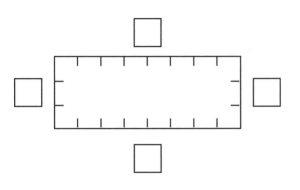

The perimeter is _____ inches.

2. The shape is a:

square   circle   triangle

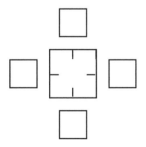

The perimeter is _____ inches.

3. The shape is a:

    square   rectangle   triangle

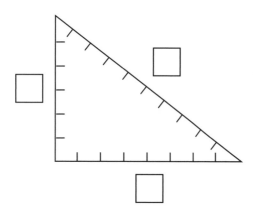

    The perimeter is _____ inches.

4. The shape is a:

    square   rectangle   triangle

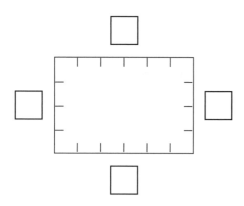

    The perimeter is _____ inches.

Circle the name of each shape. Add to find the perimeter. The lengths are given for you.

1. The shape is a:

square   circle   triangle

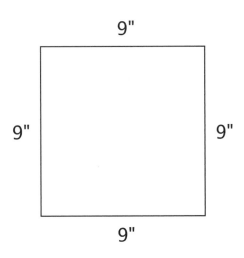

9"

9"   9"

9"

The perimeter is _____ inches.

2. The shape is a:

square   rectangle   triangle

5 in

2 in   2 in

5 in

The perimeter is _____ inches.

3. The shape is a:

square   circle   triangle

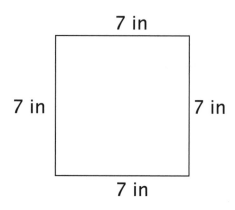

7 in

7 in   7 in

7 in

The perimeter is _____ inches.

4. The shape is a:

square   rectangle   triangle

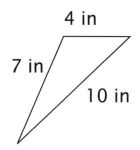

4 in

7 in

10 in

The perimeter is _____ inches.

Circle the name of each shape. Add to find the perimeter. The lengths are given for you.

1. The shape is a:

   square   rectangle   triangle

   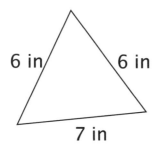

   The perimeter is _____ inches.

2. The shape is a:

   square   rectangle   triangle

   15 in

   10 in [   ] 10 in

   15 in

   The perimeter is _____ inches.

Add.

3. $2.4 9
   + 1.3 2
   ‾‾‾‾‾

4.   3 0 0
   + 4 0 9
   ‾‾‾‾‾

5.
$$\begin{array}{r} 27 \\ +\ 25 \\ \hline \end{array}$$

Review these subtraction facts.

6. 11 – 5 = ____   7. 12 – 4 = ____

8. 13 – 5 = ____   9. 11 – 3 = ____

10. 12 – 5 = ____   11. 11 – 4 = ____

12. Ashley is seven years old. How many years will go by before she is ten years old?

   _____

13. Anthony sold 53 newspapers in the morning and 8 more in the evening. How many papers did he sell in all?

   _____

14. Luke drew a line two feet long. How many inches long was the line?

   _____

15. Seth has nine nickels and one dime. How many cents does he have?

   _____

Circle the name of the shape. Add to find the perimeter.

1. The shape is a:

    square   circle   triangle

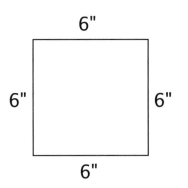

The perimeter is _____ inches.

2. The shape is a:

    square   rectangle   triangle

    12 in

    5 in   5 in

    12 in

The perimeter is _____ inches.

Add.

3. $4.2 8
    + 1.6 5

4.   2 8 5
    + 1 5 6

5.   4 5
       + 5 5

Add across and down. Then add your answers and see if they match.

6.

7	9	
3	5	

7.

6	4	
9	8	

Review these subtraction facts.

8. $14 - 5 = $ _____

9. $12 - 3 = $ _____

10. $6 - 1 = $ _____

11. $13 - 4 = $ _____

12. Ron drew a square that was four inches long on each side. What was the perimeter of the square?

_____

13. January and March each had 31 days. February had only 28 days. How many days were in those three months?

_____

14. Katie is four feet tall. How many inches tall is Katie?

_____

Circle the name of the shape. Add to find the perimeter.

1. The shape is a:

      square   rectangle   triangle

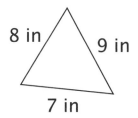

The perimeter is _____ inches.

2. The shape is a:

      square   rectangle   triangle

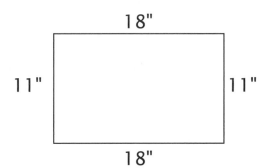

The perimeter is _____ inches.

Add.

3. $2.4 8
    + 1.2 9

4.   1 8 3
    +   7 7

5.
$$
\begin{array}{r}
6\ 3 \\
+\ 2\ 6 \\
\hline
\end{array}
$$

Skip count and write the numbers.

6. 2, ____, ____, ____, ____, ____, ____, ____, ____, ____

7. 5, ____, ____, ____, ____, ____, ____, ____, ____, ____

Review these subtraction facts.

8. $9 - 6 =$ ____

9. $11 - 7 =$ ____

10. $12 - 5 =$ ____

11. $13 - 9 =$ ____

12. Ava invited 11 people to her birthday party. Five of them have already arrived. How many guests have not yet come?

_____

13. Donna made a garden shaped like a triangle. The sides of the triangle were six feet, eight feet, and ten feet. How many feet of fence does she need to go all around her garden? (This is a perimeter problem.)

_____

14. The team had nine runs in the first inning and nine runs in the second inning. How many runs do they have so far?

_____

15G

This enrichment introduces metric units.

Most rulers have inches on one edge and centimeters on the other edge.

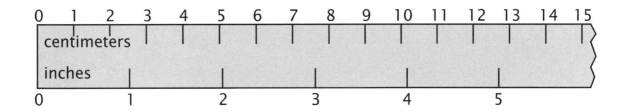

One inch is a little more than two and one-half centimeters.

Estimate the length of each line. Measure the line using inches and then using centimeters.

1. ▬▬▬▬▬▬▬▬▬▬▬▬▬▬▬▬

   My estimate is _____ inches, or _____ centimeters.

   The line is _____ inches long.

   The line is _____ centimeters long.

2. ▬▬▬▬▬▬▬▬▬

   My estimate is _____ inches, or _____ centimeters.

   The line is between _____ inches and _____ inches long.

   The line is _____ centimeters long.

To the teacher: The best way to learn measurement is by measuring real objects. Have students practice measuring objects using centimeters, inches, or feet. Encourage them to estimate before measuring.

Add the given measures to find the perimeter.

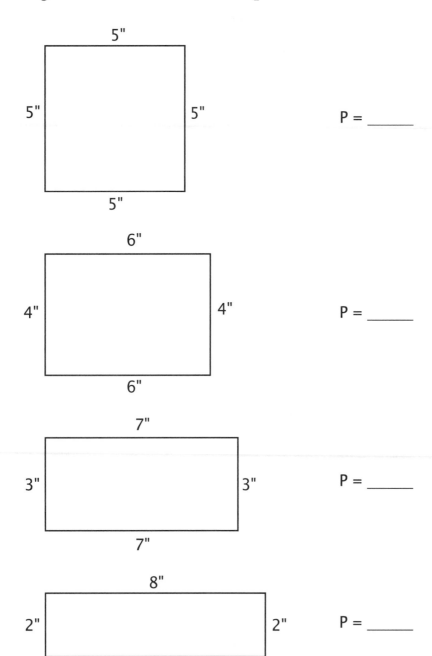

P = _____

P = _____

P = _____

P = _____

Can different rectangles have the same perimeter?

Write the number and say it. The first one has been done for you.

1. $1,000 + 400 + 10 + 8 = \underline{1,418}$

2. $30,000 + 5,000 + 300 + 60 + 1 =$ _____

3. $700,000 + 80,000 + 5,000 + 800 + 90 + 2 =$ _____

4. $200,000 + 60,000 + 5,000 + 100 + 40 + 3 =$ _____

5. $6,000 + 200 + 30 + 7 =$ _____

Write using place-value notation. The first one has been done for you.

6. $13,129 = \underline{10,000} + \underline{3,000} + \underline{100} + \underline{20} + \underline{9}$

7. $2,356 =$ _____ + _____ + _____ + _____

Write the number and say it. The first one has been done for you.

8. Three hundred forty-five thousand, four hundred eighty-two

    $\underline{345,482}$

9. One thousand, five hundred forty-two _____

Add. The answers will include the thousands place. The first one has been done for you.

10.
$$
\begin{array}{r}
{\scriptstyle 1} \\
8\,6\,3 \\
+\,3\,5\,1 \\
\hline
1,2\,1\,4
\end{array}
$$

11.
$$
\begin{array}{r}
9\,1\,5 \\
+\,4\,3\,6 \\
\hline
\end{array}
$$

12.
$$
\begin{array}{r}
3\,8\,1 \\
+\,7\,2\,7 \\
\hline
\end{array}
$$

Write the number and say it.

1. $2,000 + 700 + 90 + 4 =$ _____

2. $10,000 + 6,000 + 300 + 20 + 2 =$ _____

3. $600,000 + 50,000 + 1,000 + 700 + 40 + 1 =$ _____

4. $500,000 + 30,000 + 6,000 + 500 + 80 + 3 =$ _____

5. $2,000 + 500 + 40 + 9 =$ _____

Write using place-value notation.

6. $41,456 =$ _____ + _____ + _____ + _____ + _____

7. $238,199 =$ _____ + _____ + _____ + _____ + _____ + ____

Write the number and say it.

8. Three thousand, one hundred twenty-one _____

9. Forty-five thousand, six hundred sixteen _____

Add. The answers will include the thousands place.

10.
```
   5 9 3
 + 5 5 1
 _____
```

11.
```
   8 7 6
 + 4 3 1
 _____
```

12.
```
   9 6 7
 + 2 0 2
 _____
```

Write the number and say it.

1. 1,000 + 200 + 20 + 4 = _____

2. 40,000 + 3,000 + 600 + 30 + 8 = _____

3. 200,000 + 40,000 + 7,000 + 500 = _____

4. 100,000 + 20,000 + 2,000 + 400 + 70 + 2 = _____

5. 7,000 + 200 + 90 + 4 = _____

Write using place-value notation.

6. 56,644 = _____ + _____ + _____ + _____ + _____

7. 3,256 = _____ + _____ + _____ + _____

Write the number and say it.

8. One thousand, eight hundred thirty-eight _____

9. Thirty-three thousand, two hundred thirty _____

Add. The answers will include the thousands place.

10.
```
   4 5 3
 + 7 1 4
```

11.
```
   3 4 5
 + 9 7 8
```

12.
```
   7 1 6
 + 5 6 3
```

Write the number and say it.

1. 4,000 + 800 + 10 + 9 = _____

2. 50,000 + 7,000 + 200 + 80 + 4 = _____

Circle the name of the shape. Add to find the perimeter. Remember that *in* means inches.

3. The shape is a:

square   rectangle   triangle

32 in

11 in   11 in

32 in

The perimeter is _____ inches.

Add.

4.   9 0 1
     + 8 5 0

5.   5 3 4
     + 6 7 3

6. $4.1 2
      + 4.7 1

7.   1 7
        2 3
      + 5 5

Review these subtraction facts.

8. $10 - 6 =$ ____

9. $15 - 8 =$ ____

10. $14 - 6 =$ ____

11. $6 - 3 =$ ____

12. $12 - 7 =$ ____

13. $6 - 4 =$ ____

14. Abigail went shopping. If she spent $48 on the first day, $32 on the second day, and $21 on the third day, how much did she spend in all?

_____

15. Lisa drew a line three feet long. How many inches long was the line?

_____

16. John is 16, and Justin is 8. What is the difference in their ages?

_____

Write the number and say it.

1. 6,000 + 200 + 10 + 1 = _____

2. Twenty-eight thousand, six hundred sixteen _____

Circle the name of the shape. Add to find the perimeter.

3. The shape is a:

    square   rectangle   triangle

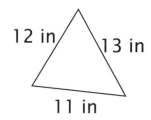

    The perimeter is _____ inches.

Add.

4.　　5 5 9
　　+ 5 2 4

5.　　9 4 3
　　+ 4 7 5

6.　$3.3 7
　+ 8.2 9

7.　　6 5
　　　2 8
　　+ 4 5

Review these subtraction facts.

8. 9 – 3 = _____

9. 8 – 5 = _____

10. 14 – 8 = _____

11. 11 – 4 = _____

12. 16 – 7 = _____

13. 7 – 1 = _____

14. Molly bought two pieces of furniture for her new house. They cost $198 and $242. Round to the nearest hundred and estimate. Then find exactly what she spent.

_____

15. Jim had eight dimes. He gave his little brother 30¢. How much money does Jim have left?

_____

16. A garden is shaped like a rectangle. It is nine feet long and seven feet wide. How many feet of fence are needed to go all around the garden? (This is a perimeter problem.)

_____

16F

Write the number and say it.

1. 7,000 + 800 + 50 + 4 = _____

2. Eight hundred fifteen thousand, two hundred thirty-one =

_____

Circle the name of the shape. Add to find the perimeter.

3. The shape is a:

square   rectangle   triangle

The perimeter is _____ inches.

Add.

4. 6 2 4
   + 4 1 3

5. 4 2 6
   + 8 7 3

6. $1.3 2
   + 3.3 8

7. 7 1
    5 3
   + 1 1

Review these subtraction facts.

8. 8 - 7 = _____

9. 11 - 8 = _____

10. 14 - 9 = _____

11. 9 - 4 = _____

12. 13 - 5 = _____

13. 8 - 3 = _____

14. James met 512 cars on his way to work and 471 cars on his way home. How many cars did he meet that day? Estimate first, and then solve.

_____

15. Christina needs 10 apples to make a pie. If she has 7 apples now, how many more does she need?

_____

16. Which is worth more: five nickels or four dimes?

_____

You can also find the perimeter of shapes that are not simple rectangles or triangles.

What is the perimeter of this house? Pretend you are an ant walking around the outside of the house and add up all the sides of the house. Remember that *ft* means feet.

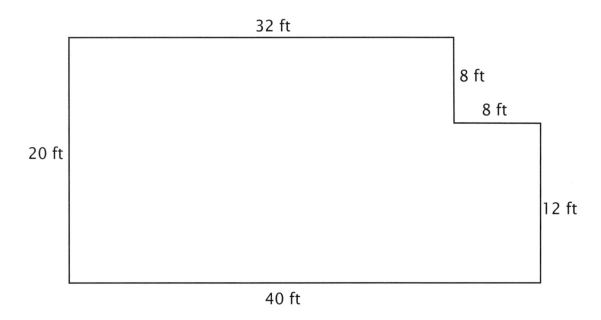

The perimeter of the house is _____ feet.

If you wish, you may draw lines inside the house to show where you think the different rooms should be.

Help to write the word problems. Fill in names of people you know.
Write your answers in the boxes.

1. _____ ran 10 miles on Monday.
On Tuesday she walked 12 miles. On Wednesday she rode
her bike for 15 miles. How far did she travel altogether?

2. _____ painted two pictures in May,
eight pictures in June, and seven pictures in July. In August,
nine pictures were sold. How many pictures are left?

3. _____ went skydiving. Last year
115 jumps were made, and this year 237 jumps were made.
How many jumps did the skydiver make in all?

Round to the nearest thousand.

1. 4,850 → _____

2. 1,324 → _____

Round to the nearest ten thousand.

3. 87,416 → _____

4. 52,933 → _____

Round to the nearest thousand and estimate the answer. Then find the exact answer. The first one has been done for you.

5.
$$\begin{array}{r} \overset{1\ 1}{7,3\ 7\ 3} \\ +\ 4,6\ 6\ 1 \\ \hline 1\ 2,0\ 3\ 4 \end{array}$$
$$\begin{array}{r} (7,000) \\ +\ (5,000) \\ \hline (12,000) \end{array}$$

6.
$$\begin{array}{r} 4,8\ 5\ 9 \\ +\ 2,4\ 4\ 4 \\ \hline \end{array}$$
$$\begin{array}{r} (\quad) \\ +\ (\quad) \\ \hline (\quad) \end{array}$$

7.
$$\begin{array}{r} 9,2\ 5\ 3 \\ +\ 7,8\ 4\ 5 \\ \hline \end{array}$$
$$\begin{array}{r} (\quad) \\ +\ (\quad) \\ \hline (\quad) \end{array}$$

8.
$$\begin{array}{r} 7,1\ 3\ 2 \\ +\ 1,1\ 8\ 6 \\ \hline \end{array}$$
$$\begin{array}{r} (\quad) \\ +\ (\quad) \\ \hline (\quad) \end{array}$$

9. 3,6 2 4   (   )
    + 4,4 1 8   + (   )
                 (   )

10. 2,8 5 2   (   )
     + 3,1 4 9   + (   )
                  (   )

11. Sue traveled 3,152 miles this week and 7,321 miles last week. How far did she travel in all?

    _____

12. There were 5,232 big fish and 3,765 little fish that swam up the stream. How many fish were there altogether?

    _____

Round to the nearest thousand.

1. 7,056 → _____

2. 3,512 → _____

Round to the nearest ten thousand.

3. 45,346 → _____

4. 21,918 → _____

Round to the nearest thousand and estimate the answer. Then find the exact answer.

5. $\begin{array}{r} 5,242 \\ +\ 3,765 \\ \hline \end{array}$ $\begin{array}{r} (\quad\quad) \\ +\ (\quad\quad) \\ \hline (\quad\quad) \end{array}$
6. $\begin{array}{r} 9,287 \\ +\ 1,321 \\ \hline \end{array}$ $\begin{array}{r} (\quad\quad) \\ +\ (\quad\quad) \\ \hline (\quad\quad) \end{array}$

7. $\begin{array}{r} 6,463 \\ +\ 8,765 \\ \hline \end{array}$ $\begin{array}{r} (\quad\quad) \\ +\ (\quad\quad) \\ \hline (\quad\quad) \end{array}$
8. $\begin{array}{r} 7,214 \\ +\ 1,108 \\ \hline \end{array}$ $\begin{array}{r} (\quad\quad) \\ +\ (\quad\quad) \\ \hline (\quad\quad) \end{array}$

9.
$$
\begin{array}{r}
2,817 \\
+\ 9,236 \\
\hline
\end{array}
$$
(   )
+ (   )
  (   )

10.
$$
\begin{array}{r}
3,680 \\
+\ 3,384 \\
\hline
\end{array}
$$
(   )
+ (   )
  (   )

11. Christina planted 5,740 carrot seeds and 4,291 celery seeds. How many seeds did she plant?

_____

12. Debbie counted 4,987 stars in one part of the sky and 3,732 stars in another part of the sky. How many stars did Debbie count altogether?

_____

Round to the nearest thousand.

1. 1,476 → _____

2. 5,746 → _____

Round to the nearest ten thousand.

3. 79,813 → _____

4. 14,375 → _____

Round to the nearest thousand and estimate the answer. Then find the exact answer.

5.   9, 4 1 3   (   )
    + 1, 2 4 5   + (   )
    _____   (   )

6.   9, 2 8 7   (   )
    + 7, 4 9 1   + (   )
    _____   (   )

7.   5, 4 8 6   (   )
    + 4, 5 2 8   + (   )
    _____   (   )

8.   9, 0 2 5   (   )
    + 3, 3 5 4   + (   )
    _____   (   )

9. $\begin{array}{r} 7,513 \\ + 7,254 \\ \hline \end{array}$ (   )
    + (   )
    (   )

10. $\begin{array}{r} 1,890 \\ + 3,672 \\ \hline \end{array}$ (   )
    + (   )
    (   )

11. Susan bought a used car for $5,486. Then she spent $1,194 on repairs. How much did she spend altogether on her car?

    _____

12. There are 8,972 people living in Greenville. Last night 9,221 more people came to watch the parade. How many people were in Greenville last night?

    _____

Round to the nearest thousand.

1. 3,189 → _____

2. 1,492 → _____

Add.

3. 6, 7 8 8
  + 2, 4 6 7

4. 2, 3 5 5
  + 1, 6 7 2

5. $8.4 2
  + 3.2 1

Write the number and say it.

6. 3,000 + 100 + 80 + 8 = _____

Add to find the perimeter. Remember that 8" means 8 inches.

7. Perimeter = _____

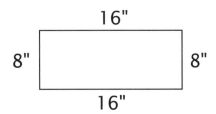

Review these subtraction facts.

8. 14 – 7 = _____   9. 11 – 6 = _____

10. 7 – 3 = _____   11. 12 – 3 = _____

12. 9 – 5 = _____   13. 8 – 4 = _____

14. Danny picked up five eggs in each hand. Then he dropped three eggs. How many whole eggs are left? Be careful!

_____

15. The first time Katherine shot her arrow, it went 296 feet. The second time she shot it, the arrow flew 316 feet. How many feet did the arrow fly altogether?

_____

16. Tom spent $16 for a shirt, $18 for a pair of pants, and $9 for a tie. How much did he spend on his new clothes?

_____

Round to the nearest ten thousand.

1. 36,254 → _____

2. 84,531 → _____

Add.

3.
```
  1, 4 7 6
+ 7, 8 1 3
```

4.
```
  1, 6 2 1
+ 4, 1 5 7
```

5.
```
  $7.1 6
+ 2.7 9
```

Write the number and say it.

6. Five thousand, four hundred _____

Add to find the perimeter. Remember that 15' means 15 feet.

7. Perimeter = _____

Review these subtraction facts.

8. 12 – 4 = _____

9. 10 – 3 = _____

10. 7 – 4 = _____

11. 3 – 2 = _____

12. 15 – 6 = _____

13. 9 – 3 = _____

14. Corey has three dimes and four pennies. How much money does he have?

_____

15. Greg has 75 pressed flowers and 108 dried leaves in his collection. How many items is that?

_____

16. Roy counted 1,575 mosquitoes at his picnic. Ron counted 1,892 mosquitoes at his family's picnic. How many mosquitoes in all were at the two picnics?

_____

Round to the nearest thousand.

1. 1,206 → _____

2. 8,738 → _____

Add.

3.   7,4 3 8
    + 9,1 1 4

4.   6,4 0 8
    + 4,3 7 9

5. $2.5 6
    + 1.2 4

Write the number and say it.

6. One hundred thirty-one thousand, five hundred twenty-eight

_____

Add to find the perimeter.

7. Perimeter = _____

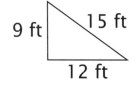

Review these subtraction facts.

8. 13 – 4 = ____

9. 11 – 3 = ____

10. 14 – 5 = ____

11. 10 – 9 = ____

12. 6 – 0 = ____

13. 8 – 8 = ____

14. Twelve people came to Madison's birthday party. Seven of them won prizes. How many did not win prizes?

_____

15. Kitty drove 235 miles yesterday and 256 miles today. Estimate how far she drove. Then find the exact answer.

_____

16. Paula gathered flowers in her garden. She cut 10 roses, 12 lilies, and 18 daisies. How many flowers are in her bouquet?

_____

Jacob likes rockets, so he made his garden in the shape of a rocket.

What is the perimeter of Jacob's garden? Pretend you are a rabbit hopping around the outside of the garden and add up all the sides of the garden.

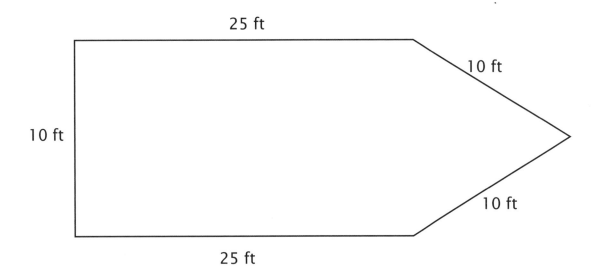

The perimeter of the garden is _____ feet.

If you wish, you may color the garden to show what kinds of vegetables or flowers should be planted in it.

Help to write the word problems. Fill in the blanks with names of insects or other creatures. Write your answers in the boxes.

1. The explorer found 1,234 _____

   and 3,452 _____ in the jungle.

   How many creatures did the explorer find?

   _____

Fill in the blanks with names of fruit.

2. Emma had fruit trees in her yard. This year she picked

   2,125 _____ and 6,741 _____.

   How many fruits did she pick altogether?

   _____

Fill in the blanks with what ever you wish.

3. One night Jayden dreamed about 134 _____

   and 7,680 _____.

   How many things did Jayden dream about?

   _____

Round to the nearest hundred and estimate. Then find the exact answer. The lines are to help you keep your place. The first one has been done for you.

1.
```
  ¹4 ⁷7 6        ( 5 0 0 )
   3  3 5        ( 3 0 0 )
   2  2 5        ( 2 0 0 )
 +  4  0 1    +  ( 4 0 0 )
 1, 4  3 7       ( 1 , 4 0 0 )
```

2.
```
   2 9 4     (      )
   1 8 7     (      )
   3 0 6     (      )
 + 8 1 3     (      )
          +  (      )
             (      )
```

3.
```
   4 9 3     (      )
   2 1 5     (      )
   4 8 5     (      )
   3 2 4     (      )
 + 1 0 6     (      )
          +  (      )
             (      )
```

4.*
```
   6 1 3     (       )
     9 7     (  1 0 0 )
   4 5 2     (       )
   8 7 9     (       )
 +   3 0  +  (     0 )
             (       )
```

*The numbers in a column will not always add to 10. In the first column of this problem, we have 7 + 3 = 10 and 9 + 2 = 11. If necessary, 10 + 11 can be written to the side and added together before continuing with the rest of the problem.

5.
$$
\begin{array}{r}
1\,|\,1\,|\,3 \\
2\,|\,5\,|\,1 \\
3\,|\,4\,|\,5 \\
3\,|\,5\,|\,5 \\
+\ 4\,|\,2\,|\,7 \\
\end{array}
$$
( )
( )
( )
( )
+ ( )
( )

6.
$$
\begin{array}{r}
5\,|\,4\,|\,6 \\
1\,|\,2\,|\,0 \\
3\,|\,0\,|\,9 \\
6\,|\,7\,|\,5 \\
+\ 4\,|\,8\,|\,1 \\
\end{array}
$$
( )
( )
( )
( )
+ ( )
( )

If you wish, turn a sheet of notebook paper sideways and use the lines to keep your numbers lined up.

7. Mrs. Brown fixed up her new house. She spent $127 on paint, $475 for a new chair, and $225 for curtains. First estimate how much money she spent. Then find the exact answer.

_____

8. Nancy drove to visit her sister. She drove 390 miles the first day, 240 miles the second day, and 152 miles the third day. About how many miles did she drive (estimate)? What is the exact number of miles she drove?

_____

Round to the nearest hundred and estimate. Then find the exact answer. Remember to look for 10.

1.
```
  2|6|4     (        )
    8|5      (        )
  6|2|4     (        )
+ 9|4|5   + (        )
          _____
            (        )
```

2.
```
  1|7|2     (        )
  2|6|1     (        )
  5|2|7     (        )
+ 4|4|6   + (        )
          _____
            (        )
```

3.
```
  9|3|3     (        )
    5|8      (        )
  3|6|1     (        )
  1|5|9     (        )
+ 5|4|2   + (        )
          _____
            (        )
```

4.
```
  1|4|2     (        )
  2|0|6     (        )
  8|6|0     (        )
  4|6|2     (        )
+ 5|5|3   + (        )
          _____
            (        )
```

5.
```
  3|2|1     (        )
    3|9      (        )
  6|8|6     (        )
  4|5|2     (        )
+ 1|5|2   + (        )
          _____
            (        )
```

6.
```
  2|1|4     (        )
  5|9|6     (        )
  4|7|3     (        )
  5|2|7     (        )
+ 8|0|2   + (        )
          _____
            (        )
```

7. In one year, Burton bought 208 gallons of gasoline for his little car, 316 gallons for his big car, and 365 gallons for his pickup truck. How many gallons of gasoline did Burton buy?

   _____

8. In Jane's school there are 137 first graders, 122 second graders, 101 third graders, and 150 fourth graders. How many students are in those four grades?

   _____

Round to the nearest hundred and estimate. Then find the exact answer. Remember to look for 10.

1.
```
    6 4 9      (    )
    5 3 6      (    )
      3 1      (    )
  + 2 2 4    + (    )
             ─────────
               (    )
```

2.
```
    7 1 4      (    )
    7 4 6      (    )
    4 1 9      (    )
  + 6 5 2    + (    )
             ─────────
               (    )
```

3.
```
    3 2 8      (    )
    5 3 0      (    )
    3 5 6      (    )
    4 3 2      (    )
  + 4 5 6    + (    )
             ─────────
               (    )
```

4.
```
    2 1 9      (    )
    8 2 0      (    )
    4 7 9      (    )
    3 8 1      (    )
  + 1 6 1    + (    )
             ─────────
               (    )
```

5.
```
    5 6 2      (    )
    5 2 0      (    )
      3 8      (    )
    6 5 9      (    )
  + 6 1 3    + (    )
             ─────────
               (    )
```

6.
```
    2 4 7      (    )
    2 5 4      (    )
    4 1 6      (    )
    2 5 2      (    )
  + 5 4 7    + (    )
             ─────────
               (    )
```

7. Julia was bored, so she counted the flakes in a box of cereal. She found that there were 553 itty-bitty ones, 334 medium-sized ones, and 129 large ones. How many flakes were in the box of cereal?

_____

8. Shirley's collection of soda cans included 302 root beer cans, 148 ginger ale cans, and 447 other cans. How many soda cans did Shirley have?

_____

Add.

1.
```
  2 9 6
  3 0 8
  7 4 2
  8 6 6
+ 3 1 4
```

2.
```
  5 2 1
    5 2
  6 2 4
  5 4 6
+   3 8
```

3.
```
  2 7 3
  9 5 1
  5 5 0
  3 3 9
+ 2 8 2
```

Write the number and say it.

4. 7,000 + 800 + 20 + 1 = _____

5. Three hundred fifty thousand _____

Review these subtraction facts.

6.
```
  2
- 0
```

7.
```
  4
- 1
```

8. $\begin{array}{r} 6 \\ -\ 2 \\ \hline \end{array}$ 9. $\begin{array}{r} 5 \\ -\ 5 \\ \hline \end{array}$

Write the correct sign in each blank. The first two have been done for you.

10. 5 $-$ 3 = 2   11. 6 $+$ 1 = 7

12. 10 ___ 8 = 2   13. 2 ___ 5 = 7

Skip count and write the numbers.

14. ____, 4, ____, 8, ____, ____, ____, ____, ____, ____

15. At Rachel's wedding, 51 guests were very early, and 29 guests were just a little early. The other 19 guests were just in time. How many guests did she have?

_____

16. As Tess walked through the meadow, she noticed that there were 68 red flowers, 89 blue flowers, and 132 purple flowers. How many flowers did she notice?

_____

Add.

1.
```
  4 5 7
  5 6 1
  4 5 1
  6 3 1
+   2 9
```

2.
```
  8 7 3
  2 6 5
  3 1 4
  2 4 7
+ 9 3 6
```

3.
```
  8 0 4
  2 4 6
  5 3 1
  3 8 2
+ 6 3 9
```

Write the number and say it.

4. 80,000 + 1,000 + 200 + 40 + 6 = _____

5. Six hundred fifty-two thousand, six hundred ninety-three

_____

Review these subtraction facts.

6. $\begin{array}{r} 1\,2 \\ -\ 9 \\ \hline \end{array}$ 7. $\begin{array}{r} 1\,5 \\ -\ 9 \\ \hline \end{array}$

8. $\begin{array}{r} 1\,1 \\ -\ 9 \\ \hline \end{array}$ 9. $\begin{array}{r} 1\,7 \\ -\ 9 \\ \hline \end{array}$

Write the correct sign in each blank.

10. $9 \underline{} 4 = 13$ 11. $18 \underline{} 9 = 9$

12. $14 \underline{} 9 = 5$ 13. $9 \underline{} 7 = 16$

Skip count and write the numbers.

14. ____, ____, 15, 20, ____, ____, ____, ____, ____, ____

15. The tree in Madison's yard was six feet tall. How many inches tall was her tree?

    _____

16. Bailey found three nickels and four dimes. How much money did she find?

    _____

Add.

1.
$$
\begin{array}{r}
7\ 4\ 1 \\
4\ 0\ 5 \\
4\ 0\ 1 \\
5\ 8\ 6 \\
+\ 3\ 6\ 4 \\
\end{array}
$$

2.
$$
\begin{array}{r}
9\ 5\ 2 \\
2\ 3\ 7 \\
1\ 7\ 1 \\
6\ 0\ 3 \\
+\ 4\ 5\ 9 \\
\end{array}
$$

3.
$$
\begin{array}{r}
6\ 3\ 1 \\
8\ 3\ 4 \\
8\ 4\ 1 \\
2\ 7\ 2 \\
+\ 2\ 3\ 4 \\
\end{array}
$$

Write the number and say it.

4. $600,000 + 30,000 + 7,000 + 500 + 30 + 1 =$

   _____

5. Forty-five thousand, seven hundred twenty-seven

   _____

Review these subtraction facts.

6. $\begin{array}{r} 11 \\ -\ 8 \\ \hline \end{array}$

7. $\begin{array}{r} 14 \\ -\ 8 \\ \hline \end{array}$

8. $\begin{array}{r} 17 \\ -\ 8 \\ \hline \end{array}$

9. $\begin{array}{r} 13 \\ -\ 8 \\ \hline \end{array}$

Write the correct sign in each blank.

10. 15 _ 8 = 7

11. 12 _ 8 = 4

12. 8 _ 8 = 16

13. 9 _ 8 = 1

Skip count and write the numbers.

14. ____, 20, ____, ____, 50, ____, ____, ____, ____, ____

15. Liane read three books. The first had 87 pages, the next had 48 pages, and the last had 211 pages. How many pages did Liane read?

_____

16. Caleb hopes to get his license when he is 16 years old. If he is 7 years old now, how many years does he have to wait?

_____

Use a yardstick or meter stick to find these measurements. (A meter is the same as 100 centimeters.) Remember to write inches, centimeters, or meters after each answer.

1. Measure a table with your yardstick or meter stick.

    How long is the table?_____

    How wide is the table? _____

2. Measure a rug with your yardstick or meter stick.

    How long is the rug?_____

    How wide is the rug? _____

3. Measure a window with your yardstick or meter stick.

    How tall is the window?_____

    How wide is the window? _____

4. How tall are you? Get someone in your family to help you.

    _____

Note: Continue to round the answers to the nearest inch or centimeter. If students want to measure longer lengths, you can introduce them to a tape measure.

Now you can find the perimeters of the things you measured on the last enrichment sheet.

1. Draw a picture of the table you measured and write the numbers to show how long and how wide it is.

2. What is the perimeter of the table? _____

3. Draw a picture of the rug you measured and write the numbers to show how long and how wide it is.

4. What is the perimeter of the rug? _____

5. Draw a picture of the window you measured and write the numbers to show how long and how wide it is.

6. What is the perimeter of the window? _____

19A

Round to the nearest thousand and estimate the answer. Then find the exact answer. Remember to look for 10. The first one has been done for you.

1.
```
    1 1
  1,5 8 2     ( 2,0 0 0 )
  3,6 2 4     ( 4,0 0 0 )
 + 4 1 3    + (   0 0 0 )
  5,6 1 9     ( 6,0 0 0 )
```

2.
```
  7,1 3 2     (       )
  5,3 3 3     (       )
 + 1,1 8 6  + (       )
              (       )
```

3.
```
  2,8 5 2     (       )
  4,2 6 3     (       )
 + 3,1 4 9  + (       )
              (       )
```

4.
```
  6,7 3 2     (       )
  3,1 5 2     (       )
 + 7,3 2 1  + (       )
              (       )
```

5.
```
    5,2 3 2    (        )
    7,1 1 1    (        )
  + 3,7 6 5  + (        )
              (        )
```

6.
```
    1,2 5 7    (        )
    6,4 6 3    (        )
  + 8,7 6 5  + (        )
              (        )
```

You may turn a sheet of notebook paper sideways and use the lines to keep your columns lined up.

7. Claire traveled from Boston to Atlanta (1,083 miles). Then she traveled to Tampa (482 miles). Next she traveled to Detroit (1,216 miles). Finally Claire traveled back to Boston (741 miles). How many miles did Claire travel altogether?

_____

8. For Claire's next adventure, she flew to San Francisco (3,187 miles) and then on to Los Angeles (408 miles). From there she flew to Dallas (1,443 miles). After that she flew home to Boston (1,837 miles). How many miles did she fly altogether?

_____

19B

Round to the nearest thousand first and estimate the answer. Then find the exact answer. Remember to look for 10.

1.
$$\begin{array}{r} 2,8\,1\,7 \\ 9,2\,3\,6 \\ +\ 3,6\,8\,0 \\ \hline \end{array}$$

$$\begin{array}{r} (\qquad) \\ (\qquad) \\ +\ (\qquad) \\ \hline (\qquad) \end{array}$$

2.
$$\begin{array}{r} 5,7\,4\,0 \\ 4,2\,2\,1 \\ +\ 3,3\,2\,1 \\ \hline \end{array}$$

$$\begin{array}{r} (\qquad) \\ (\qquad) \\ +\ (\qquad) \\ \hline (\qquad) \end{array}$$

3.
$$\begin{array}{r} 3,2\,1\,3 \\ 1,3\,5\,7 \\ +\ 2,7\,9\,8 \\ \hline \end{array}$$

$$\begin{array}{r} (\qquad) \\ (\qquad) \\ +\ (\qquad) \\ \hline (\qquad) \end{array}$$

4.
$$\begin{array}{r} 1,4\,7\,6 \\ 7\,4\,6 \\ +\ 9,8\,1\,3 \\ \hline \end{array}$$

$$\begin{array}{r} (\qquad) \\ (\qquad) \\ +\ (\qquad) \\ \hline (\qquad) \end{array}$$

5.
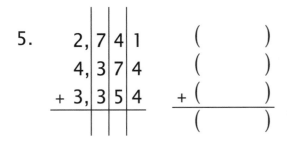

$$\begin{array}{r} 2,741 \\ 4,374 \\ + \ 3,354 \\ \hline \end{array}$$

(   )
(   )
+ (   )
(   )

6.
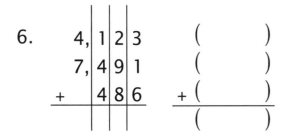

$$\begin{array}{r} 4,123 \\ 7,491 \\ + \ \ \ 486 \\ \hline \end{array}$$

(   )
(   )
+ (   )
(   )

7. The fisherman caught 2,365 fish the first day, 1,295 fish the second day, and 3,116 the third day. How many fish did he catch in all?

_____

8. In Pineville 7,851 students go to school, and 3,895 are homeschooled. How many students live in Pineville?

_____

Round to the nearest thousand first and estimate the answer. Then find the exact answer. Remember to look for 10.

1.
```
    3,6 9 5        (    )
    3,1 7 5        (    )
  + 2,1 4 1     + (    )
                   (    )
```

2.
```
    1,4 6 8        (    )
    6,0 1 2        (    )
  + 5,2 8 0     + (    )
                   (    )
```

3.
```
    2,9 4 0        (    )
    4,2 7 8        (    )
  + 1,9 6 3     + (    )
                   (    )
```

4.
```
    1,0 8 6        (    )
    3,6 0 8        (    )
  + 2,6 5 7     + (    )
                   (    )
```

5.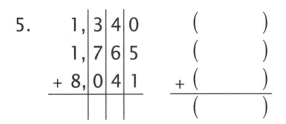

6.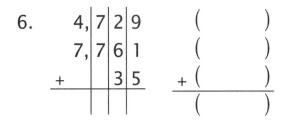

7. Smithville has 8,716 people living in it, and Jonesville has 6,658 people. What is the total number of people who live in the two towns?

   _____

8. Mr. White has a furniture store. On Monday he had three customers who spent $2,345, $5,634, and $1,954. How much did they spend in all?

   _____

19D

Add.

1.
```
  2,4|7|5
  1,8|9|0
+   |3|7|6
```

2.
```
  7,5|1|3
  9,0|2|5
+ 3,2|5|4
```

3.
```
  3,1|8|9
  1,4|2|2
+ 2,4|6|8
```

Review these subtraction facts. We will begin to learn multiple-digit subtraction in the next lesson.

4.
```
   8
 − 4
```

5.
```
  1 0
 −  6
```

6.
```
  1 4
 −  7
```

7.
```
  1 0
 −  9
```

8.
```
   6
 − 3
```

9.
```
  1 0
 −  3
```

10.
$$\begin{array}{r} 4 \\ -\ 2 \\ \hline \end{array}$$

11.
$$\begin{array}{r} 1\ 0 \\ -\ 5 \\ \hline \end{array}$$

Compare. Then fill in the oval with <, >, or =.

12. $12 - 6 \bigcirc 2 + 2$

13. $10 + 4 \bigcirc 10 - 4$

14. $10 - 7 \bigcirc 10 - 3$

15. Jason earned $35 on Monday, $42 on Tuesday, $33 on Wednesday, and $45 on Thursday. How much money did Jason earn altogether?

_____

16. Hannah needs $17 to buy a gift for her mother. She already has $8. How much more money does she need?

_____

17. Seth is raising earthworms to sell. One box has 3,645 earthworms, one has 4,782, and the last has 5,641. How many earthworms does Seth have to sell?

_____

18. Alexis drew a triangle. Each side was five inches long. What was the perimeter of her triangle?

_____

Add.

1.
$$\begin{array}{r} 2{,}3\,8\,4 \\ 4{,}1\,2\,3 \\ +\,6{,}3\,3\,5 \\ \hline \end{array}$$

2.
$$\begin{array}{r} 3{,}5\,9\,1 \\ 2{,}3\,6\,7 \\ +\,8{,}4\,5\,9 \\ \hline \end{array}$$

3.
$$\begin{array}{r} 2{,}3\,6\,8 \\ 4{,}1\,5\,2 \\ +\,6{,}3\,1\,4 \\ \hline \end{array}$$

Review these subtraction facts. We will begin to learn multiple-digit subtraction in the next lesson.

4.
$$\begin{array}{r} 9 \\ -\,5 \\ \hline \end{array}$$

5.
$$\begin{array}{r} 2 \\ -\,1 \\ \hline \end{array}$$

6.
$$\begin{array}{r} 9 \\ -\,6 \\ \hline \end{array}$$

7.
$$\begin{array}{r} 10 \\ -\,2 \\ \hline \end{array}$$

8.
$$\begin{array}{r} 9 \\ -\,3 \\ \hline \end{array}$$

9.
$$\begin{array}{r} 16 \\ -\,9 \\ \hline \end{array}$$

10.   9
    − 4

11.   9
    − 7

Compare. Then fill in the oval with <, >, or =.

12. 12 − 3 $\bigcirc$ 12 + 3

13. 16 − 7 $\bigcirc$ 14 − 5

14. 6 − 4 $\bigcirc$ 9 − 1

15. Forrest bought a birthday card that cost $2.98 and a balloon that cost $3.75. How much did he spend in all?

_____

16. Paula found seven dimes. She lost five of them. How many cents does she have left?

_____

17. A builder is ordering cement. For different jobs he needs 20 tons, 18 tons, 10 tons, and 35 tons. How many tons of cement should he order?

_____

18. Mr. Brown's vegetable garden is shaped like a square. Each side of the square is 35 feet long. How many feet of fence does Mr. Brown need to go all around his garden?

_____

Add.

1.
$$
\begin{array}{r}
8,4\,8\,2 \\
4,6\,2\,1 \\
+\ 5,3\,5\,1 \\
\hline
\end{array}
$$

2.
$$
\begin{array}{r}
1,2\,6\,4 \\
7,6\,3\,2 \\
+\ 1,9\,5\,3 \\
\hline
\end{array}
$$

3.
$$
\begin{array}{r}
5,1\,4\,8 \\
2,6\,3\,3 \\
+\ 4,1\,8\,6 \\
\hline
\end{array}
$$

Review these subtraction facts. Be sure you know your subtraction facts before going to the next lesson.

4.
$$
\begin{array}{r}
7 \\
-\ 3 \\
\hline
\end{array}
$$

5.
$$
\begin{array}{r}
1\,1 \\
-\ 6 \\
\hline
\end{array}
$$

6.
$$
\begin{array}{r}
7 \\
-\ 4 \\
\hline
\end{array}
$$

7.
$$
\begin{array}{r}
1\,3 \\
-\ 5 \\
\hline
\end{array}
$$

8.
$$
\begin{array}{r}
8 \\
-\ 3 \\
\hline
\end{array}
$$

9.
$$
\begin{array}{r}
9 \\
-\ 2 \\
\hline
\end{array}
$$

10.   8
     − 5
     ‾‾‾

11.   1 1
     − 4
     ‾‾‾

Compare. Then fill in the oval with <, >, or =.

12. 8 − 6 $\bigcirc$ 8 − 7

13. 12 − 7 $\bigcirc$ 12 + 7

14. 14 − 6 $\bigcirc$ 4 + 4

15. A truck driver drove the following numbers of miles on different days: 495, 382, 516, and 402. How many miles did he drive in all?

_____

16. Dave's yard is a rectangle. Two sides are 25 feet long, and two sides are 35 feet long. What is the perimeter of his yard?

_____

17. Aunt Dot knitted mittens for six people. If each person has two hands, how many mittens did Aunt Dot need to make altogether? (skip count)

_____

18. Four of the mittens Aunt Dot knitted were red, and the rest were blue. Use your answer for number 17 and find out how many mittens were blue.

_____

Add.

If the answer is 100, color the space yellow.
If the answer is 200, color the space black or gray.
If the answer is 300, color the space brown.
If the answer is 400, color the space blue.
If the answer is 500, color the space red.

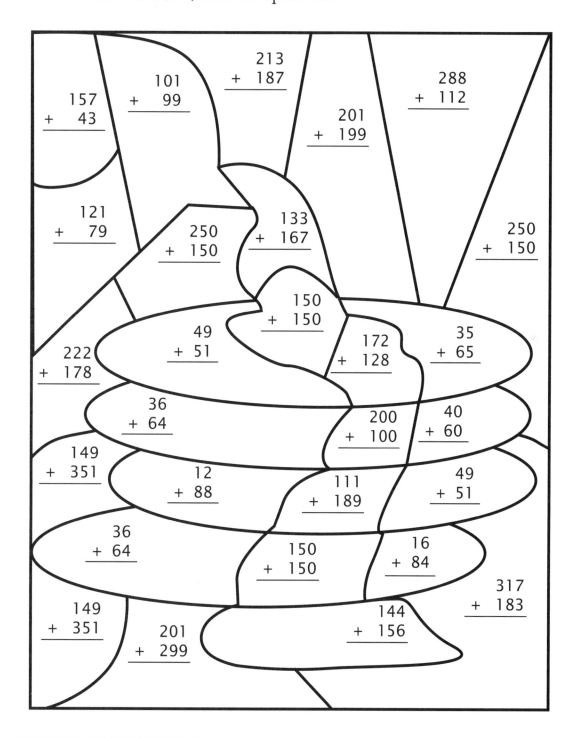

Subtraction Facts: Mark the ones you know. Practice until you know them all.

0 – 0	1 – 1	2 – 2	3 – 3	4 – 4	5 – 5	6 – 6	7 – 7	8 – 8	9 – 9
1 – 0	2 – 1	3 – 2	4 – 3	5 – 4	6 – 5	7 – 6	8 – 7	9 – 8	10 – 9
2 – 0	3 – 1	4 – 2	5 – 3	6 – 4	7 – 5	8 – 6	9 – 7	10 – 8	11 – 9
3 – 0	4 – 1	5 – 2	6 – 3	7 – 4	8 – 5	9 – 6	10 – 7	11 – 8	12 – 9
4 – 0	5 – 1	6 – 2	7 – 3	8 – 4	9 – 5	10 – 6	11 – 7	12 – 8	13 – 9
5 – 0	6 – 1	7 – 2	8 – 3	9 – 4	10 – 5	11 – 6	12 – 7	13 – 8	14 – 9
6 – 0	7 – 1	8 – 2	9 – 3	10 – 4	11 – 5	12 – 6	13 – 7	14 – 8	15 – 9
7 – 0	8 – 1	9 – 2	10 – 3	11 – 4	12 – 5	13 – 6	14 – 7	15 – 8	16 – 8
8 – 0	9 – 1	10 – 2	11 – 3	12 – 4	13 – 5	14 – 6	15 – 7	16 – 8	17 – 9
9 – 0	10 – 1	11 – 2	12 – 3	13 – 4	14 – 5	15 – 6	16 – 7	17 – 8	18 – 9

20A

Subtract and check by adding. The first one has been done for you. (Choose the way of checking you like best. You don't need to check the problem twice!)

1. $\begin{array}{r} 7\ 2 \\ -\ 3\ 1 \\ \hline 4\ 1 \\ \hline 7\ 2 \end{array}$ check: $\begin{array}{r} 3\ 1 \\ +\ 4\ 1 \\ \hline 7\ 2 \end{array}$

2. $\begin{array}{r} 6\ 0 \\ -\ 4\ 0 \\ \hline \end{array}$

3. $\begin{array}{r} 9\ 4 \\ -\ 5\ 1 \\ \hline \end{array}$

4. $\begin{array}{r} 5\ 3 \\ -\ 4\ 2 \\ \hline \end{array}$

5. $\begin{array}{r} 4\ 0 \\ -\ 3\ 0 \\ \hline \end{array}$

6. $\begin{array}{r} 4\ 5\ 9 \\ -\ 3\ 1\ 2 \\ \hline \end{array}$

7. $\begin{array}{r} 4\ 4 \\ -\ 2\ 0 \\ \hline \end{array}$

8. $\begin{array}{r} 9\ 2\ 4 \\ -\ \ \ 1\ 3 \\ \hline \end{array}$

9.
```
  5 0 6
- 3 0 2
```

10.
```
   2 5
 - 2 1
```

11.
```
  8 4 1
- 6 2 0
```

12.
```
  9 9 9
- 1 2 3
```

13. There are 32 students in Mrs. Martin's class. Ten students are out with the flu today. How many are in the class today?

_____

14. Emaleigh raised rabbits. She had 44 rabbits, but someone accidently left the cage doors open, and 11 rabbits ran away. How many rabbits are left?

_____

LESSON PRACTICE

Subtract and check by adding.

1. 35
    − 2 4

2. 26
    − 1 3

3. 50
    − 2 0

4. 83
    − 1 2

5. 49
    − 4 7

6. 989
    − 4 3 2

7. 46
    − 3 0

8. 554
    − 2 1

9.   3 0 0
       – 1 0 0

10.   6 2
       – 3 0

11.   4 3 8
       – 2 1 4

12.   3 9 7
       – 1 7 5

13. Daniel collects baseball cards. He had 68 cards, but he traded 25 of them for some special marbles. How many baseball cards did he have left?

_____

14. A truck had 48 gallons of gasoline in the tank when it started its trip. If 23 gallons were left at the end of the trip, how many gallons were used?

_____

LESSON PRACTICE

Subtract and check by adding.

1. $\begin{array}{r} 6\ 5 \\ -\ 3\ 2 \\ \hline \end{array}$

2. $\begin{array}{r} 1\ 7 \\ -\ 1\ 7 \\ \hline \end{array}$

3. $\begin{array}{r} 5\ 2 \\ -\ 2\ 1 \\ \hline \end{array}$

4. $\begin{array}{r} 2\ 0 \\ -\ 1\ 0 \\ \hline \end{array}$

5. $\begin{array}{r} 7\ 5 \\ -\ 3\ 2 \\ \hline \end{array}$

6. $\begin{array}{r} 1\ 8\ 8 \\ -\ 2\ 3 \\ \hline \end{array}$

7. $\begin{array}{r} 6\ 9 \\ -\ 3\ 1 \\ \hline \end{array}$

8. $\begin{array}{r} 5\ 6\ 1 \\ -\ 2\ 6\ 0 \\ \hline \end{array}$

9.
$$
\begin{array}{r}
6\ 4\ 5 \\
-\ 4\ 3\ 5 \\
\hline
\end{array}
$$

10.
$$
\begin{array}{r}
8\ 5 \\
-\ 4\ 4 \\
\hline
\end{array}
$$

11.
$$
\begin{array}{r}
2\ 2\ 5 \\
-\ 1\ 0\ 0 \\
\hline
\end{array}
$$

12.
$$
\begin{array}{r}
5\ 3\ 8 \\
-\ 4\ 2\ 1 \\
\hline
\end{array}
$$

13. Twenty-nine children at summer camp got prize ribbons. Thirteen children lost their ribbons. How many still had ribbons to take home?

_____

14. Chad had $49. He spent $21 on a gift for his sister. How much money does Chad have left?

_____

SYSTEMATIC REVIEW

Subtract and check by adding.

1.
```
   7 7
 - 1 1
```

2.
```
   5 0
 - 4 0
```

3.
```
   3 9
 - 2 7
```

4.
```
   6 9 3
 - 3 6 1
```

5.
```
   3 0 0
 - 1 0 0
```

6.
```
   1 6 3
 -   5 1
```

Add.

7.
```
   2,1 7 4      (          )
   7,4 1 8      (          )
 + 3,7 9 1    + (          )
              (          )
```

8.
```
   2,5 6 4      (          )
   6,4 0 8      (          )
 + 1,2 4 3    + (          )
              (          )
```

9.

3	7	9
5	1	1
3	3	3
+ 4	6	8

(   )
(   )
+ (   )
(   )

Write the number and say it.

10. $100,000 + 20,000 + 4,000 + 900 + 70 + 1 =$ _____

Skip count and write the numbers.

11. ____, ____, 6, ____, 10, ____, ____, ____, ____, ____

12. ____, 10, ____, ____, 25, ____, ____, ____, ____, ____

13. ____, ____, 30, ____, ____, ____, ____, ____, ____, 100

14. Carolyn had 25¢ in her pocket. If she spent 14¢, how much did she have left?

_____

15. During their vacation, Jason's family traveled 103 miles by car, 3,521 miles by airplane, and 58 miles by train. How many miles did they travel in all? Be sure you have each digit in the correct column when you set up the problem.

_____

Subtract and check by adding.

1.
```
   3 5
 - 2 2
```

2.
```
   4 9
 - 3 1
```

3.
```
   2 8
 -   5
```

4.
```
   6 3 3
 - 5 1 0
```

5.
```
   7 9 0
 - 6 9 0
```

6.
```
   5 6 1
 - 5 5 0
```

Add.

7.
```
   1,8 9 0      (    )
   3,6 7 2      (    )
 + 7,2 5 4    + (    )
              ──────────
                (    )
```

8.
```
   7,1 9 3      (    )
   4,6 8 5      (    )
 + 1,4 9 2    + (    )
              ──────────
                (    )
```

9.

```
    9 2 2      (      )
    6 7 8      (      )
    2 5 3      (      )
  + 1 1 2    + (      )
  ─────────  ──────────
             (      )
```

Write the number and say it.

10. Two hundred five thousand, nine hundred eighteen

_____

Skip count and write the numbers.

11. 5, ____, ____, ____, ____, ____, ____, ____, ____, ____

12. ____, 4, ____, ____, ____, 12, ____, ____, ____, ____

13. ____, ____, ____, 40, ____, ____, ____, ____, 90, ____

14. Sandi had 49 pennies. If she bought a candy bar that cost 23 cents, how much money did she have left over?

_____

15. A factory bought 285 boxes of apples, 683 boxes of peaches, and 491 boxes of pears to make juice. How many boxes of fruit were bought in all?

_____

Subtract and check by adding.

1.　　9 0
　　 – 7 0

2.　　5 1
　　 – 1 0

3.　　7 4
　　 – 3 1

4.　 1 3 9
　 – 1 2 5

5.　 9 9 9
　 – 3 4 7

6. 1 6 7
　 – 2 4

Add.

7.　2,4 6 7　　(　　)
　　9,2 2 1　　(　　)
　+ 6,7 8 8　+ (　　)
　　　　　　　 (　　)

8.　4,2 5 3　　(　　)
　　4,1 1 2　　(　　)
　+ 1,2 5 5　+ (　　)
　　　　　　　 (　　)

9.

```
  2 3 8      (        )
  4 1 2      (        )
  6 3 3      (        )
+ 4 5 9    + (        )
           ──────────
             (        )
```

Write the number and say it.

10. $20,000 + 3,000 + 600 + 80 + 4 =$ _____

Skip count and write the numbers.

11. ____, 20, ____, ____, ____, ____, 70, ____, ____, ____

12. ____, ____, 15, ____, ____, ____, 35, ____, ____, ____

13. ____, ____, ____, 8, ____, ____, 14, ____, ____, ____

14. A yardstick is three feet long. How many inches long is it?

      _____

15. Jacki counted 67 cars on the way to Grandma's house. That is 13 more than Vicki saw. Subtract to find how many cars Vicki saw.

      _____

Use the words and the clues to fill in the crossword puzzle.*

add
addends
difference
minuend
perimeter
subtract
subtrahend
sum

Across

3. The top number in a subtraction problem is called the _____ .

5. The bottom number in a subtraction problem is called the _____ .

7. Add all the sides of a shape to find the _____ .

8. To find how many in all, you should _____ .

Down

1. To find the difference between two numbers, you should _____ .

2. The answer to a subtraction problem is called the _____ .

4. The answer to an addition problem is called the _____ .

6. The numbers that are added are called _____ .

*The names for the parts of addition and subtraction problems are in Lessons 5 and 20 in the Instruction Manual.

Fill in the blanks. Subtract and circle the answer that shows how much change each person should get. The first one has been done for you.

1. A toy costs 29¢. Mia gave the clerk three dimes.

$$\begin{array}{r} 30¢ \\ -29¢ \\ \hline 1¢ \end{array}$$

Three dimes is the same as __30¢__ . Subtract and circle the picture that shows how much change Mia should get.

2. A treat costs 27¢. Joshua gave the clerk two dimes and two nickels.

_____ − _____

Two dimes and two nickels are the same as _____ . Circle the picture that shows how much change Joshua should get.

3. An ice cream cone costs 51¢. Emma gave the clerk six dimes.

_____ − _____

Six dimes is the same as _____ . Circle the picture that shows how much change Emma should get.

21A

Write the number of minutes shown by each clock. The first one has been done for you. (There is a clock template at the end of this book.)

1.

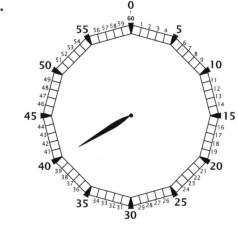

__:20__

2.

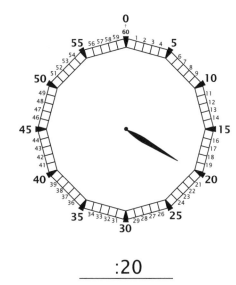

_____

3.

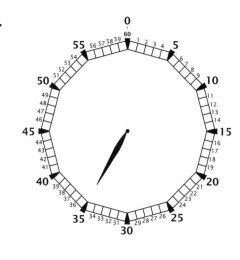

_____

4.

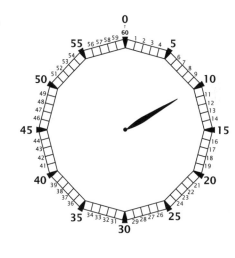

_____

5.

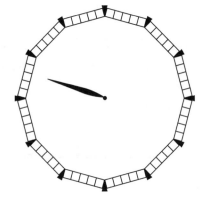

_____

6.

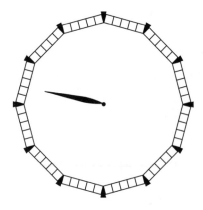

_____

Write the number of minutes shown by each clock.

1.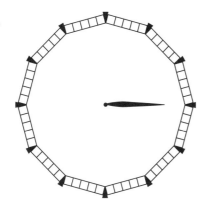

_____

2.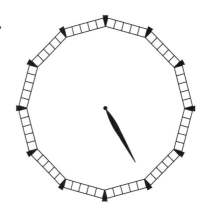

_____

3.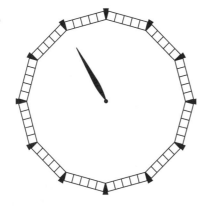

_____

4.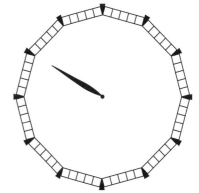

_____

5.

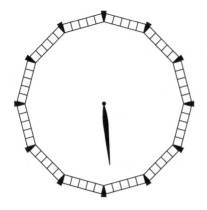

_____

6.

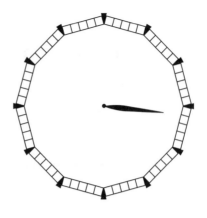

_____

21C

Write the number of minutes shown by each clock.

1.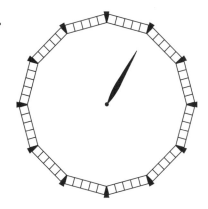

_____

2.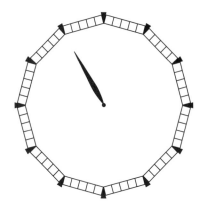

_____

3.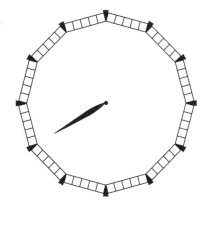

_____

4.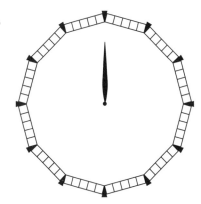

_____

5.

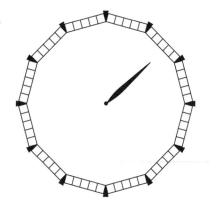

_____

6.

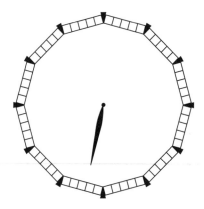

_____

Write the number of minutes shown by each clock.

1.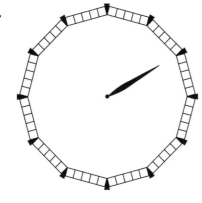

2.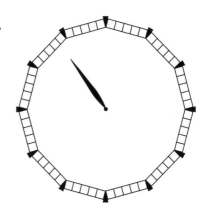

_____

_____

Subtract and check.

3. $\begin{array}{r} 99 \\ -\ 84 \\ \hline \end{array}$

4. $\begin{array}{r} 138 \\ -\ 34 \\ \hline \end{array}$

5. $\begin{array}{r} 279 \\ -\ 164 \\ \hline \end{array}$

Add.

6. $3.27
    + 4.16

7. 2,384
    + 6,335

8. 367
    591
    + 419

9. In the spring, the bugs came out to play. In John's back yard there were 4,568 gnats, 3,851 mosquitoes, and 2,001 dragonflies. How many insects were there altogether?

    _____

10. Keith is buying fencing to make a square pen for the family's pet goat. The pen will be 10 feet on each side. The gate will be 3 feet wide. How many feet of fence should Keith buy? (Be careful! This is a two-step problem.)

    _____

11. Rose is 25 years old, and Karen is 14 years old. What is the difference in their ages?

    _____

12. Dan made 19 points playing a game, and Angie made 7 points. How many more points did Dan make?

    _____

SYSTEMATIC REVIEW

Write the number of minutes shown by each clock.

1.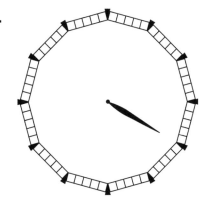

_____

2.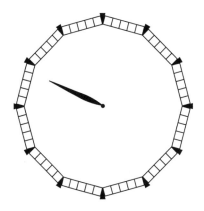

_____

Subtract and check.

3.
```
    8 3
  – 5 2
  ─────
```

4.
```
  8 7 3
  – 6 1
  ─────
```

5.
```
  3 6 2
  – 1 5 1
  ───────
```

Add.

6.
```
  $9.2 8
+  2.1 7
```

7.
```
  4,1 2 3
+ 7,4 6 0
```

8.
```
  4 5 9
  3 6 7
+ 5 9 1
```

9. Lillian is 28 years older than Pam. If Pam is 32 years old, how old is Lillian?

_____

10. Russ is 8 years younger than Jim. If Jim is 29 years old, how old is Russ?

_____

11. Greg walked all around the outside walls of the house. His house is shaped like a rectangle. It is 40 feet long and 35 feet wide. How many feet did Greg walk?

_____

12. Mom made 144 chocolate chip cookies and 120 raisin cookies. After the picnic, only 52 cookies were left over. How many cookies were eaten at the picnic?

_____

Write the number of minutes shown by each clock.

1.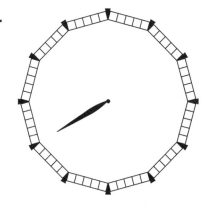

2.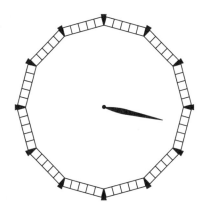

_____   _____

Subtract and check.

3.   1 9
   − 7
   _____

4. 8 3 9
  − 1 6
  _____

5.   6 0 4
  − 3 0 2
  _____

Add.

6.
```
  $5.3 6
+  1.5 1
```

7.
```
  8,4 8 2
+ 4,6 2 1
```

8.
```
  3 4 3
  2 4 6
+ 1 1 2
```

9. Shirley found four dimes and three nickels. Then she lost 25¢. How much money does she have left?

   _____

10. Simon drew a triangle with a perimeter of 18 inches. The first side was 6 inches long, and the second side was 6 inches long. What was the length of the third side of Simon's triangle? $6 + 6 + X = 18$ inches

    _____

11. April was very bored, so she started counting leaves on the bushes in the back yard. She counted the following numbers of leaves: 969, 1,345, 5,002, and 2,061. How many leaves did she count in all?

    _____

12. Clyde needs $25 to buy a gift for his mom. If he has $10, how many more dollars does he need?

    _____

21G

Draw a minute hand to show the minutes written under each clock.

:15   :20

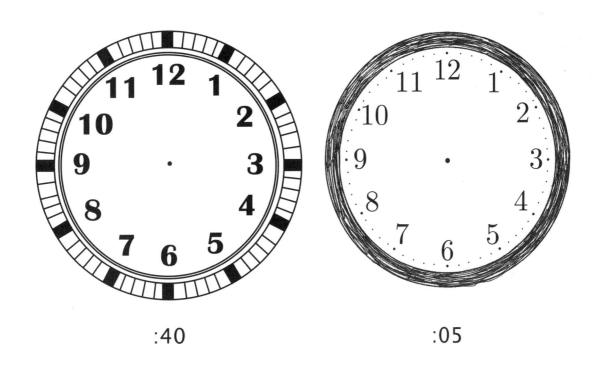

:40   :05

Draw a minute hand to show the minutes written under each clock.

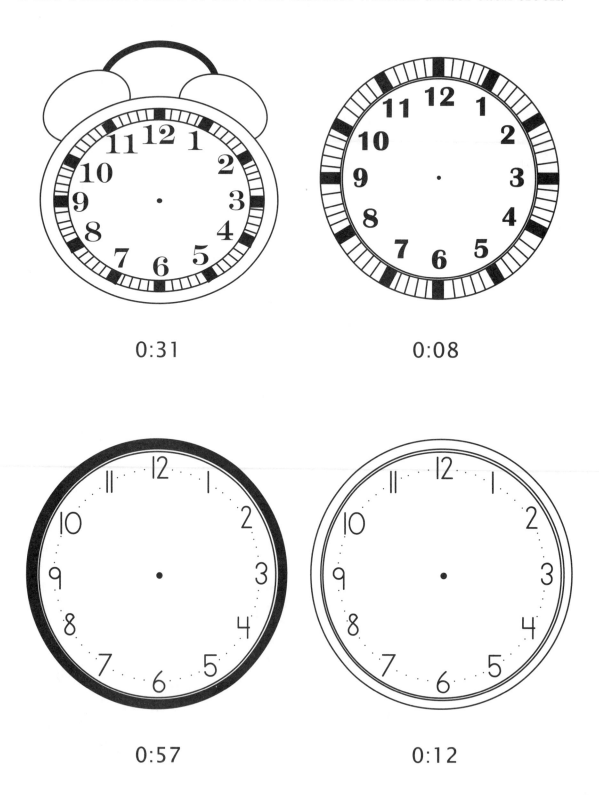

0:31   0:08

0:57   0:12

Subtract using regrouping. Check by adding. Use the method you like best. The first one has been done for you.

1.
$$\begin{array}{r} \overset{3}{\cancel{4}}\,{}^{1}2 \\ -\ 1\ 8 \\ \hline 2\ 4 \\ \hline 4\ 2 \end{array}$$
check:
$$\begin{array}{r} 1\ 8 \\ +\ 2\ 4 \\ \hline 4\ 2 \end{array}$$

2.
$$\begin{array}{r} 6\ 1 \\ -\ 2\ 7 \\ \hline \end{array}$$

3.
$$\begin{array}{r} 2\ 2 \\ -\ 1\ 3 \\ \hline \end{array}$$

4.
$$\begin{array}{r} 2\ 3 \\ -\ 1\ 9 \\ \hline \end{array}$$

5.
$$\begin{array}{r} 5\ 5 \\ -\ 2\ 6 \\ \hline \end{array}$$

6.
$$\begin{array}{r} 3\ 2 \\ -\ 1\ 4 \\ \hline \end{array}$$

7.
$$\begin{array}{r} 7\ 3 \\ -\ 3\ 4 \\ \hline \end{array}$$

8.
$$\begin{array}{r} 6\ 8 \\ -\ 2\ 9 \\ \hline \end{array}$$

9.
$$\begin{array}{r} 6\ 3 \\ -\ 4\ 9 \\ \hline \end{array}$$

10. Kelsey had 75¢. She paid 57¢ for a yo-yo. How much money did she have left?

   _____

11. Briley made 43 fancy hair clips to sell at a craft fair. If she sold 28 of them, how many did she have left?

   _____

12. An elephant ate 65 peanuts on Tuesday and 72 peanuts on Thursday. How many more peanuts did it eat on Thursday?

   _____

22B

Subtract using regrouping. Check by adding.

1.
$$\begin{array}{r} 57 \\ -29 \\ \hline \end{array}$$

2.
$$\begin{array}{r} 30 \\ -18 \\ \hline \end{array}$$

3.
$$\begin{array}{r} 65 \\ -47 \\ \hline \end{array}$$

4.
$$\begin{array}{r} 52 \\ -14 \\ \hline \end{array}$$

5.
$$\begin{array}{r} 85 \\ -49 \\ \hline \end{array}$$

6.
$$\begin{array}{r} 72 \\ -27 \\ \hline \end{array}$$

7.
$$\begin{array}{r} 53 \\ -18 \\ \hline \end{array}$$

8.
$$\begin{array}{r} 31 \\ -16 \\ \hline \end{array}$$

9.
$$\begin{array}{r} 82 \\ -53 \\ \hline \end{array}$$

10. Wayne's book has 43 pages. He has read 24 pages. How many pages are left to read?

   _____

11. Stan tossed the basketball at the hoop 61 times. He missed the hoop 36 times. How many times did the ball go through the hoop?

   _____

12. Kitty has 50 cents. How much money would she have left if she bought a 23-cent candy bar?

   _____

Subtract using regrouping. Check by adding.

1. $\begin{array}{r} 52 \\ -\ 25 \\ \hline \end{array}$

2. $\begin{array}{r} 71 \\ -\ \ 3 \\ \hline \end{array}$

3. $\begin{array}{r} 34 \\ -\ 15 \\ \hline \end{array}$

4. $\begin{array}{r} 87 \\ -\ \ 8 \\ \hline \end{array}$

5. $\begin{array}{r} 62 \\ -\ 27 \\ \hline \end{array}$

6. $\begin{array}{r} 23 \\ -\ 14 \\ \hline \end{array}$

7. $\begin{array}{r} 45 \\ -\ 26 \\ \hline \end{array}$

8. $\begin{array}{r} 31 \\ -\ 29 \\ \hline \end{array}$

9. $\begin{array}{r} 72 \\ -\ 38 \\ \hline \end{array}$

Some subtraction problems give the difference and ask you to find the number that you would subtract to find that difference. You are being asked to solve for the unknown. Remember that $8 - 5 = 3$ and $8 - 3 = 5$ are related problems. The first one has been done for you.

10. Hannah had 65 pennies, but she lost some of them. If she has 59 pennies left, how many did she lose?

   $\underline{65 - X = 59 \text{ pennies}}$

   We can rewrite this as $65 - 59 = X$.
   The answer is $X = 6$ pennies

11. The difference between Sue's and Richard's ages is 19 years. Sue is older than Richard. If Sue is 48 years old, how old is Richard?

   _____

12. Stacia picked 63 apples from her apple tree. She gave some apples away and had 39 apples left for baking. How many apples did she give away?

   _____

Subtract, using regrouping when needed. Check by adding.

1.
$$\begin{array}{r} 71 \\ -43 \\ \hline \end{array}$$

2.
$$\begin{array}{r} 98 \\ -89 \\ \hline \end{array}$$

3.
$$\begin{array}{r} 35 \\ -17 \\ \hline \end{array}$$

4.
$$\begin{array}{r} 85 \\ -45 \\ \hline \end{array}$$

5.
$$\begin{array}{r} 429 \\ -311 \\ \hline \end{array}$$

6.
$$\begin{array}{r} 148 \\ -26 \\ \hline \end{array}$$

Add.

7.
$$\begin{array}{r} \$7.33 \\ +1.83 \\ \hline \end{array}$$

8.
$$\begin{array}{r} 5,263 \\ +7,554 \\ \hline \end{array}$$

9.
$$\begin{array}{r} 892 \\ 413 \\ +476 \\ \hline \end{array}$$

Write the number of minutes shown by each clock.

10.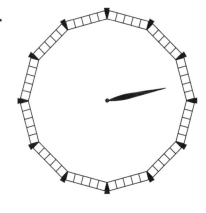

_____

11.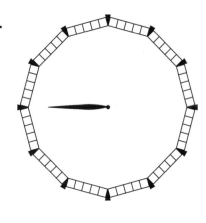

_____

12. Anthony collected 16 red marbles, 21 green marbles, and 14 blue marbles. He gave 25 of his marbles to Jack. How many marbles did Anthony have left?

_____

Subtract, using regrouping when needed. Check by adding.

1.
```
   9 2
 - 7 8
```

2.
```
   5 3
 - 4 5
```

3.
```
   2 3
 - 1 4
```

4.
```
   7 5
 - 5 0
```

5.
```
  7 4 3
 -  1 2
```

6.
```
  5 1 4
 -    2
```

Add.

7.
```
  $3.5 5
 + 2.5 2
```

8.
```
   9,4 3 5
 + 4,2 5 2
```

9.
```
   5 7 3
   8 8 2
   2 3 5
 + 7 6 2
```

Write the number of minutes shown by each clock.

10.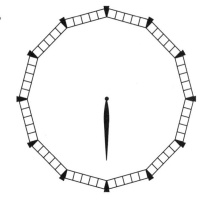

_____

11.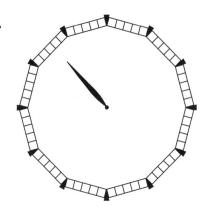

_____

12. Esther read 19 pages in her book yesterday and 21 pages today. Her book has a total of 52 pages. How many pages does she have left to read?

_____

Subtract, using regrouping when needed. Check by adding.

1.
```
   44
 - 28
```

2.
```
   93
 - 16
```

3.
```
   84
 - 27
```

4.
```
   17
 - 10
```

5.
```
   913
 - 112
```

6.
```
   307
 - 205
```

Add.

7.
```
   $6.24
 +  5.48
```

8.
```
   3,548
   7,624
 + 3,642
```

9.
```
   573
   882
 + 762
```

Write the number of minutes shown by each clock.

10.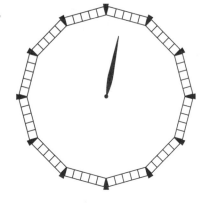

_____

11.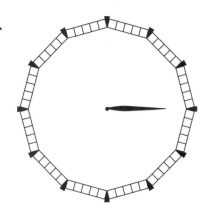

_____

12. Ranger Bill counted 134 deer, 48 elk, and 2 mountain lions. Ranger Tom counted 100 deer and 22 elk. How many more animals did Ranger Bill count?

_____

The number nine can be used to make many interesting patterns. Study the addition problems and the boxes that are filled in. Do the last addition problem and see if you can fill in the empty boxes.

$$\begin{array}{r} 10 \\ +\ 9 \\ \hline 19 \end{array} \rightarrow \quad 1 + 9 = \boxed{10}$$

$$\begin{array}{r} 100 \\ +\ 99 \\ \hline 199 \end{array} \rightarrow \quad \boxed{\begin{array}{r} 1 \\ 9 \\ +\ 9 \\ \hline 19 \end{array}} \rightarrow \quad 1 + 9 = \boxed{10}$$

$$\begin{array}{r} 1{,}000 \\ +\ 999 \\ \hline 1{,}999 \end{array} \rightarrow \quad \boxed{\begin{array}{r} 1 \\ 9 \\ 9 \\ +\ 9 \\ \hline 28 \end{array}} \rightarrow \quad 2 + 8 = \boxed{10}$$

$$\begin{array}{r} 10{,}000 \\ +\ 9{,}999 \\ \hline \end{array} \rightarrow \quad \boxed{\begin{array}{r} \\ \\ +\ \\ \hline \end{array}} \rightarrow \quad \underline{} + \underline{} = \boxed{}$$

Challenge: Add 100,000 and 99,999 and see if the answer follows the same pattern.

Help to write the word problems by choosing the numbers. The numbers should all be hundreds or thousands. Write your answers in the boxes.

1. Alex collected _____ oak leaves and _____ maple leaves. How many leaves did he collect altogether?

    ┌─────────────────────────┐
    │   │
    │   │
    │   │
    └─────────────────────────┘

2. Karen earned _____ dollars last year and _____ dollars this year. How many dollars did she earn altogether?

    ┌─────────────────────────┐
    │   │
    │   │
    │   │
    └─────────────────────────┘

Fill in the blanks. First write what Seth saw. Then choose numbers that are hundreds or thousands to tell how many he saw. Add and write your answer in the box.

3. Seth saw _____ .

    He saw _____ of them in the woods and _____ of them on the road.

    Then he saw _____ of them in his back yard. How many of them did Seth see altogether?

    ┌─────────────────────────┐
    │   │
    │   │
    │   │
    └─────────────────────────┘

What is the hour? The first one has been done for you.

1. ___9:00___

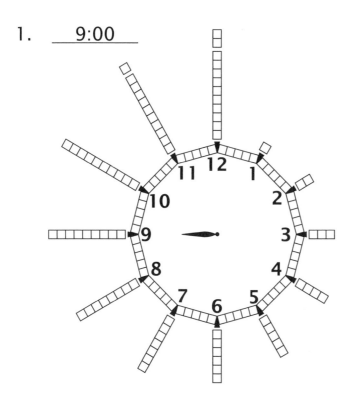

2. _____

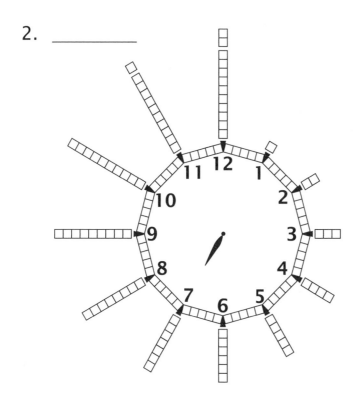

3. _____

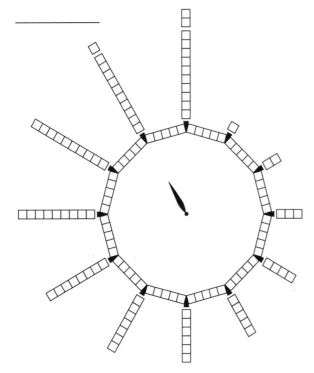

4. _____

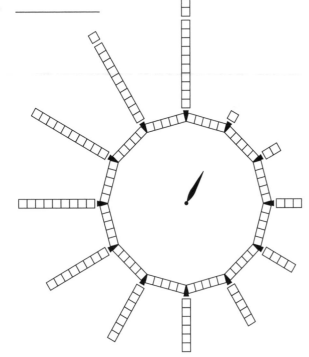

23B

What is the hour?

1. _____

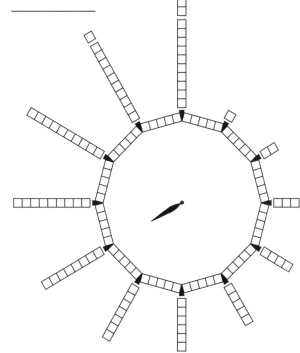

2. _____

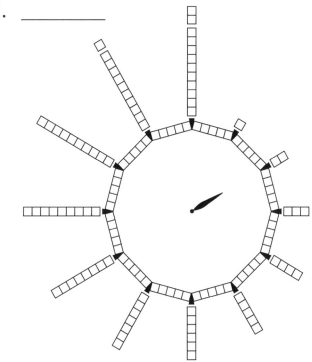

3. _____

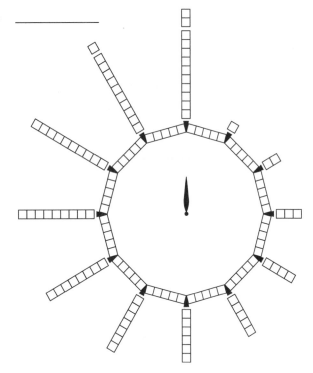

4. _____

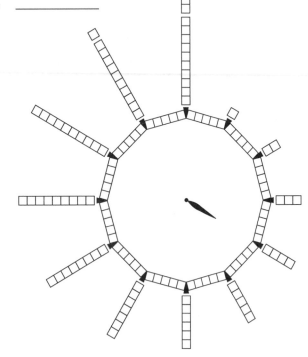

Give the time with hours and minutes. The first one has been done for you.

1.   __11:30__

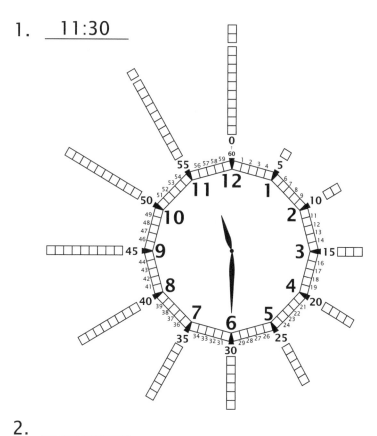

2.   _____

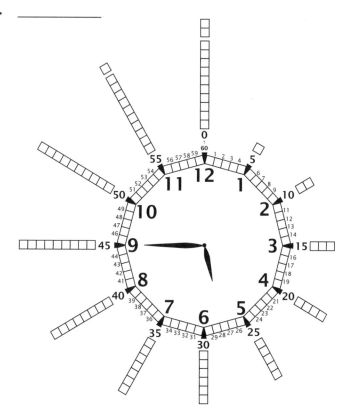

3. _____

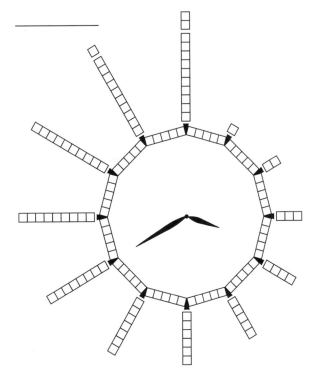

4. _____

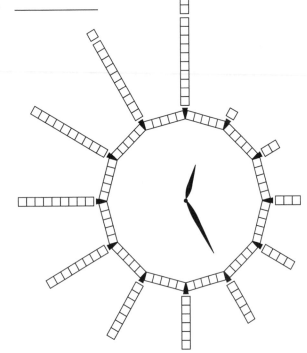

Give the time with hours and minutes.

1.

_____

2.

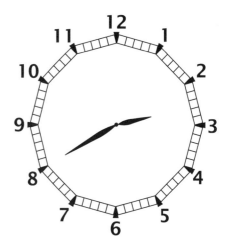

_____

3.

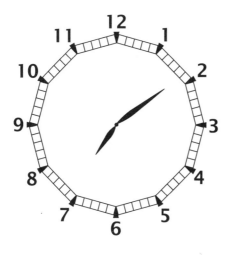

_____

4.

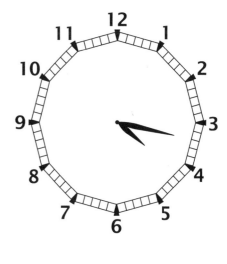

_____

Subtract and check.

5.
```
    4 0
  - 3 8
```

6.
```
    7 6
  - 1 8
```

7.
```
    6 2
  - 5 7
```

8.
```
    7 3
  - 1 4
```

9. Sonia has four dimes, seven nickels, and eight pennies. How much money does she have?

_____

10. Each side of a square field is 1,342 feet long. What is the perimeter of the field?

_____

Give the time with hours and minutes.

1.

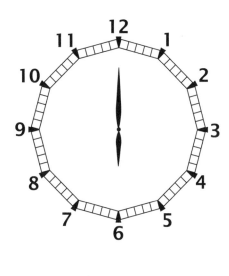

_____

2.

_____

3.

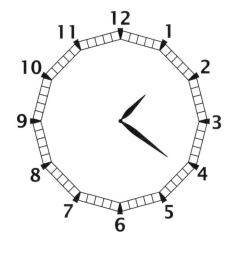

_____

4.

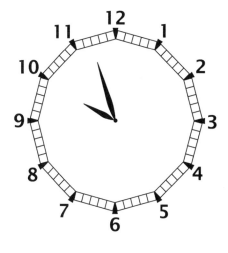

_____

Subtract and check.

5.　　7 3
　　 − 1 5

6.　　8 8
　　 − 4 4

7.　　7 6
　　 − 2 8

8.　　9 2 8
　　 − 6 0 5

9. Oliver is six feet tall. How many inches tall is he?

_____

10. Joseph had 250 toy soldiers. He got 75 more toy soldiers from Johnny for his birthday. How many toy soldiers does Joseph have now?

_____

Give the time with hours and minutes.

1.

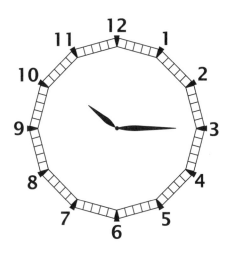

_____

2.

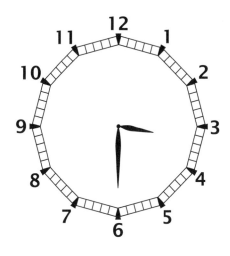

_____

3.

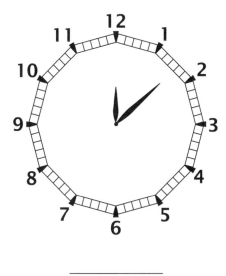

_____

4.

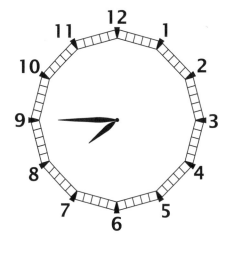

_____

Subtract and check.

5.
$$\begin{array}{r} 54 \\ -45 \\ \hline \end{array}$$

6.
$$\begin{array}{r} 39 \\ -20 \\ \hline \end{array}$$

7.
$$\begin{array}{r} 40 \\ -35 \\ \hline \end{array}$$

8.
$$\begin{array}{r} 779 \\ -314 \\ \hline \end{array}$$

9. Kelly bought a salad for $4.52 and a sandwich for $2.91. How much did Kelly spend in all?

_____

10. Thirty-five birds landed in Sandra's back yard. Then twenty-six of them flew away. How many birds were left?

_____

Draw the hour hand to show the hours written under each clock.

3:00

10:00

6:00

1:00

Draw both hands to show the time written under each clock.

5:30   12:15

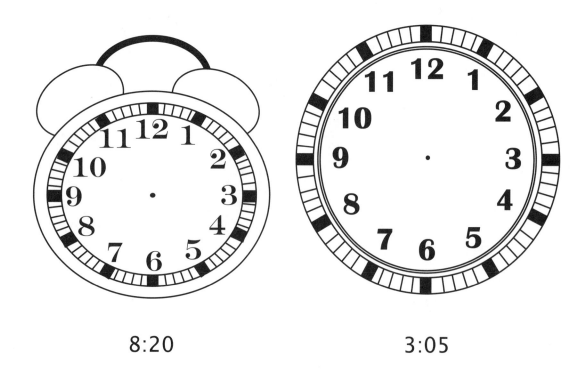

8:20   3:05

24A

Subtract using regrouping. Check by adding. The first two have been done for you.

1.
$$\begin{array}{r} \overset{1}{\cancel{2}}\overset{9}{\cancel{0}}\overset{1}{0} \\ -\ \ 2\ 5 \\ \hline 1\ 7\ 5 \\ \hline 2\ 0\ 0 \end{array}$$
check:
$$\begin{array}{r} 2\ 5 \\ +1\ 7\ 5 \\ \hline 2\ 0\ 0 \end{array}$$

2.
$$\begin{array}{r} 8\ \overset{2}{\cancel{3}}\overset{1}{5} \\ -2\ 2\ 8 \\ \hline 6\ 0\ 7 \\ \hline 8\ 3\ 5 \end{array}$$
check:
$$\begin{array}{r} 2\ 2\ 8 \\ +6\ 0\ 7 \\ \hline 8\ 3\ 5 \end{array}$$

3.
$$\begin{array}{r} 2\ 2\ 3 \\ -\ \ 8\ 7 \\ \hline \end{array}$$

4.
$$\begin{array}{r} 7\ 3\ 4 \\ -\ \ 3\ 6 \\ \hline \end{array}$$

5.
$$\begin{array}{r} 5\ 2\ 3 \\ -1\ 3\ 8 \\ \hline \end{array}$$

6.
$$\begin{array}{r} 4\ 0\ 0 \\ -3\ 9\ 9 \\ \hline \end{array}$$

7.
$$\begin{array}{r} 3\ 2\ 1 \\ -\ \ 3\ 9 \\ \hline \end{array}$$

8.
$$\begin{array}{r} 4\ 5\ 9 \\ -2\ 4\ 1 \\ \hline \end{array}$$

9.
$$\begin{array}{r} 6\ 1\ 4 \\ -1\ 9\ 6 \\ \hline \end{array}$$

10. Michael's book has 235 pages. He has read 167 pages. How many pages does he still have to read?

_____

11. Alexandria collected 300 pennies. Her brother collected 245 pennies. How many more pennies does Alexandria have than her brother has?

_____

12. While on vacation, Riley saw many tall buildings. One building was 425 feet tall, and another was 580 feet tall. What is the difference in the heights of the buildings?

_____

LESSON PRACTICE

Subtract using regrouping. Check by adding.

1. $\begin{array}{r} 716 \\ -\ 58 \\ \hline \end{array}$

2. $\begin{array}{r} 668 \\ -\ 284 \\ \hline \end{array}$

3. $\begin{array}{r} 163 \\ -\ 72 \\ \hline \end{array}$

4. $\begin{array}{r} 641 \\ -\ 49 \\ \hline \end{array}$

5. $\begin{array}{r} 372 \\ -\ 106 \\ \hline \end{array}$

6. $\begin{array}{r} 987 \\ -\ 789 \\ \hline \end{array}$

7. $\begin{array}{r} 705 \\ -\ 60 \\ \hline \end{array}$

8. $\begin{array}{r} 500 \\ -\ 250 \\ \hline \end{array}$

9. $\begin{array}{r} 471 \\ -\ 106 \\ \hline \end{array}$

10. Laura had $275 to spend on clothes. If she buys a coat that costs $116, how much does she have left to spend?

_____

11. A squirrel collected 553 acorns and buried them last fall. He dug up and ate 378 of them during the winter. How many acorns are left in the ground?

_____

12. Grace Joy planned a 325-mile trip. The first day she traveled 185 miles. How far did she have left to travel?

_____

Subtract using regrouping. Check by adding.

1. 583
    − 64

2. 203
    − 192

3. 200
    − 98

4. 319
    − 30

5. 631
    − 429

6. 419
    − 138

7. 103
    − 25

8. 572
    − 390

9. 333
    − 144

10. Eight hundred seventy-five people visited the new store. Only eighty people went home without buying something. How many people bought something in the new store?

_____

11. Joshua collected 200 baseball cards. He gave 115 of them to his friend. How many baseball cards does Joshua have now?

_____

12. Isabella bought 375 yards of yarn for a knitting project. When she had finished knitting, she had 18 yards left over. How many yards of yarn did she use?

_____

Subtract using regrouping as needed. Check by adding.

1.
```
  1 0 0
-   7 5
```

2.
```
  9 0 8
- 2 9 1
```

3.
```
  2 5 6
- 1 3 8
```

4.
```
  1 9 9
-   8 2
```

5.
```
  6 3
- 2 4
```

6.
```
  4 1
- 3 5
```

Give the time with hours and minutes.

7.

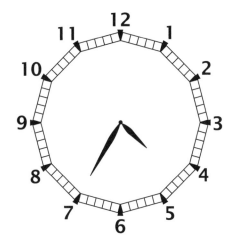

_____

8.

_____

Add.

9.
```
  4 3 6
+ 1 2 2
```

10.
```
  2 9 8
+ 5 3 9
```

11.
```
  2,9 9 9
+ 3,1 1 1
```

12. Jayden caught 175 fireflies last night and put them in his room. This morning he counted only 98 fireflies. How many are missing?

_____

13. We drove 212 miles this morning and 362 miles this afternoon. How many miles did we drive today?

_____

14. Mom had $400 to spend on furniture. She bought a chair for $215 and an end table for $134. How much did she have left over?

_____

Subtract using regrouping as needed. Check by adding.

1.　　 1 5 2
　　　－　 8 7

2.　　 3 2 0
　　　－1 1 8

3.　　 8 0 3
　　　－3 0 0

4.　　 1 9 3
　　　－　6 7

5.　　 4 3
　　　－1 7

6.　　 6 4
　　　－3 8

Give the time with hours and minutes.

7.
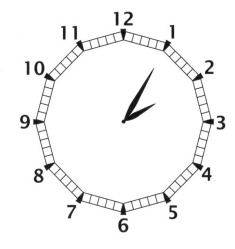

_____

8.
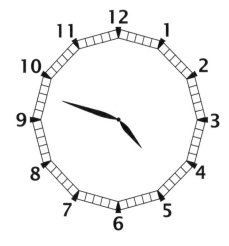

_____

Add.

9.
```
  2 1 2
+ 3 6 2
```

10.
```
  3 5 6
+ 4 8 1
```

11.
```
  1, 2 7 6
+ 7, 3 9 1
```

12. Janna read a book with 321 pages. Cassie read a book with 215 pages. What is the difference in the number of pages they read?

_____

13. Farmer John had 665 cows and 133 calves in the barn. How many animals did he have in the barn?

_____

14. Mr. Herr sold 658 ice cream cones yesterday. He sold 314 chocolate cones and 219 vanilla cones. The rest were strawberry. How many strawberry-flavored ice cream cones did he sell?

_____

Subtract using regrouping as needed. Check by adding.

1.
```
  1 0 0
-   4 7
```

2.
```
  1 4 1
- 1 1 3
```

3.
```
  2 2 0
- 1 6 4
```

4.
```
  1 5 0
-   9 8
```

5.
```
  7 0
- 2 0
```

6.
```
  3 2
- 1 8
```

Give the time with hours and minutes.

7.
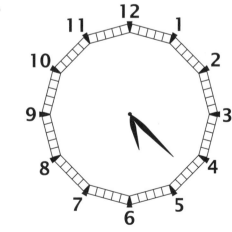

_____

8.
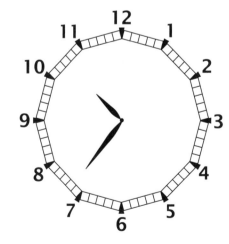

_____

Add.

9.
```
   4 0 0
 + 3 7 6
```

10.
```
   2 8 8
 + 1 3 7
```

11.
```
   2,1 4 6
 + 1,4 0 8
```

12. Five hundred eighty-one people went to the concert. Four hundred ninety-nine people loved the music, but the rest were unhappy. How many were unhappy?

_____

13. Robin spent $126 for groceries the first week and $132 the second week. How much did she spend in all for groceries?

_____

14. Abby collects little toy dogs. She already had 45 of them at the beginning of this year. She bought 11 more, got 5 for her birthday, and 8 for Christmas. How many more must she collect before she has 100 dogs?

_____

Here is a number pattern that you can find by subtracting nines. Subtract the first problem. Write the answer under the line and in the rectangle of the problem below it. Add the digits in the answer and write the answer in the square. The first one has been done for you.

$$\begin{array}{r} \boxed{99} \\ -\ 9 \\ \hline 90 \end{array} \rightarrow 9 + 0 = \boxed{9}$$

$\downarrow$

$$\begin{array}{r} \boxed{90} \\ -\ 9 \\ \hline \end{array} \rightarrow + = \boxed{}$$

$\downarrow$

$$\begin{array}{r} \boxed{} \\ -\ 9 \\ \hline \end{array} \rightarrow + = \boxed{}$$

$\downarrow$

$$\begin{array}{r} \boxed{} \\ -\ 9 \\ \hline \end{array} \rightarrow + = \boxed{}$$

$\downarrow$

(continued on the next page)

This is the same pattern that you started on the last page. Keep on subtracting nine and adding the digits in each answer.

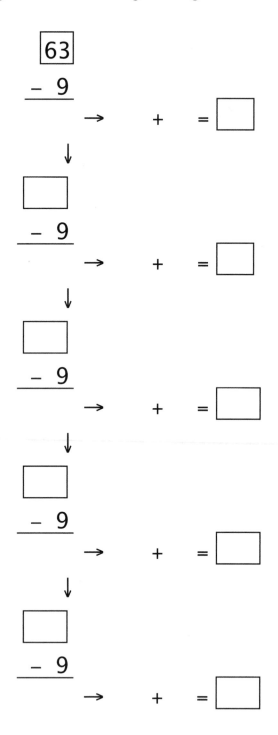

Fill in the blanks.

1. The second month of the year is _____ .

2. The _____ day of the week is Monday.

Matching.

3. first   Tuesday

4. second   Sunday

5. third   Friday

6. fourth   Thursday

7. fifth   Wednesday

8. sixth   Saturday

9. seventh   Monday

How many days are in each month?

10. January _____   11. December _____

12. June _____

Write how many.

13. I̶N̶ _____   14. III _____

15. I̶N̶ III _____

Show the number with tally marks.

16. 19   _____

17. 30   _____

18. 29   _____

19. Aaron had the hiccups and kept track of them for an hour. Show his tally marks if he hiccuped 35 times.

_____

20. The first day of summer is in June. Which month of the year is that?

_____

Fill in the blanks.

1. The fourth day of the week is _____ .

2. The _____ day of the week is Saturday.

3. The tenth month of the year is _____ .

Matching.

4. first   April

5. second   February

6. third   June

7. fourth   January

8. fifth   March

9. sixth   May

How many days are in each month?

10. February _____   11. August _____

12. November _____

Write how many.

13. ||||| ||||| | _____   14. ||||| | _____

15. ||||| ||||| ||||| _____

Show the number with tally marks.

16. 10   _____

17. 13   _____

18. 7   _____

19. Alex kept track of all the red cars he saw on the way to town. His tally looked like this: ||||| ||||| ||||| ||. How many red cars did he see?

_____

20. Which day is the first day of the week?

_____

Fill in the blanks.

1. The fourth month of the year is _____ .

2. The _____ day of the week is Tuesday.

3. The _____ day of the week is Sunday.

Matching.

4. seventh   September

5. eighth   December

6. ninth   July

7. tenth   November

8. eleventh   October

9. twelfth   August

How many days are in each month?

10. April _____   11. March _____

12. July _____

Write how many.

13. ||||| ||||| || _____   14. ||||| ||| _____

15. ||||| ||||| |||| _____

Show the number with tally marks.

16. 25 _____

17. 9 _____

18. 4 _____

19. Steve Demme was born on the fifth day of the week. What day is that?

_____

20. Name the holiday that is on the fourteenth day of the second month.

_____

Fill in the blanks.

1. The first month of the year is _____ .

2. The _____ day of the week is Wednesday.

3. The _____ day of the week is Friday.

4. May has _____ days.

Add or subtract.

5.
```
    3 5 5
  -   2 6
```

6.
```
    6 1 8
  - 2 2 4
```

7.
```
    4 1 9
  - 3 5 7
```

8.
```
    3,6 2 9
  + 2,4 2 8
```

Give the time with hours and minutes.

9.

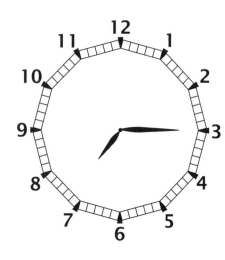

_____

10.

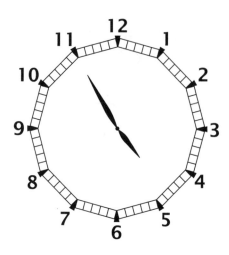

_____

11. Bill kept track of the number of times he had to erase mistakes on his math worksheet. Monday looked like this: ₩₩ ₩₩ ll; and Tuesday looked like this: ₩₩ lll. How many times did he have to erase in two days?

_____

12. Ryan had 5 red jelly beans, 17 green jelly beans, and 14 blue jelly beans. Find how many he had in all. Write your answer with tally marks.

_____

13. Bria counted 131 days until the end of school. After 47 days had gone by, how many were still left?

_____

14. What is the perimeter of a square that measures 11 feet on each side?

_____

Fill in the blanks.

1. The third month of the year is _____ .

2. The sixth day of the week is _____ .

3. The ninth month of the year is _____ .

4. September has _____ days.

Add or subtract.

5.
```
  2 3 4
-   5 2
```

6.
```
  5 9 0
- 4 1 8
```

7.
```
  9 1 0
-   7 5
```

8.
```
  1,8 3 1
+ 5,4 3 9
```

Give the time with hours and minutes.

9.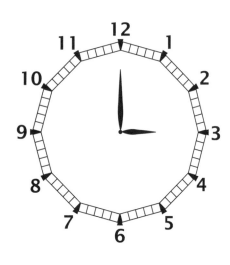

10.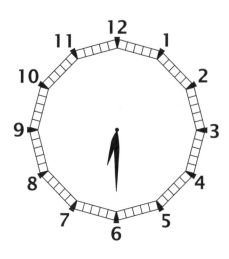

_____   _____

11. Carla kept track of how many times it rained for a month:
    IIII IIII II. How many times did it rain?

    _____

12. Use tally marks to record your age.

    _____

13. Ian flew in an airplane three times last month. The first time
    he flew 3,451 miles; the second time he flew 2,163 miles;
    and the last time he flew 1,999 miles. How many miles did
    Ian fly in all?

    _____

14. Sandi's birthday is on the first day of the eighth month. In
    what month is her birthday?

    _____

Fill in the blanks.

1. The fifth month of the year is _____ .

2. The eleventh month of the year is _____ .

3. The _____ day of the week is Friday.

4. October has _____ days.

Add or subtract.

5.
```
  5 0 7
 -  4 1
```

6.
```
  1 8 2
 - 1 4 6
```

7.
```
  9 0 0
 -  2 5
```

8.
```
  7, 1 4 7
 + 8, 7 1 3
```

Give the time with hours and minutes.

9.
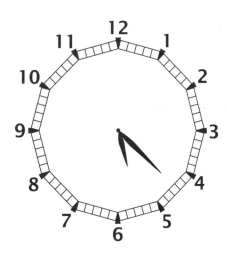

_____

10.

_____

11. Ben and David kept track of the books they read for an entire month. Ben wrote ‖‖‖ ‖‖‖ ‖‖‖ ‖‖‖, and David wrote ‖‖‖ ‖‖‖ ‖‖‖ ‖‖‖ ‖. Who read more books?

_____

12. Our family enjoys Saturdays. What day of the week is that?

_____

13. One house is 110 years old. Another house is 35 years old. What is the difference in the ages of the two houses?

_____

14. The perimeter of a triangle is 29 inches. One of the sides measures 6 inches, and another measures 10 inches. What is the length of the third side of the triangle?

_____

To the parent: There is a one-month calendar on the next application and enrichment page. Below are some ideas for its use.

A. Look at a current calendar and fill in the correct days for this month. Compare the total number of days with the list in your instruction manual.

B. Write the name of the month on the line provided. Add the ordinal number that tells what month it is.

C. Add holidays and family birthdays to the correct days.

D. If you wish, record the weather for each day. You may use words or pictures.

E. Choose two dates and subtract to find how many days are between them.

F. Write the year at the top of the calendar. Find out when family members were born and subtract to find out how old they will be this year.

G. Discuss your family's weekly schedule. Are there things that you usually do on a particular day of the week?

H. Use a current calendar that you have in your home and talk about the seasons. Discuss which months are considered spring, summer, fall, or winter months where you live.

This month is _____. It is the _____ _____ month. The year is _____.

Sunday	Monday	Tuesday	Wednesday	Thursday	Friday	Saturday

Subtract using regrouping. Check by adding in whichever way you like best. The first two have been done for you.

1.
$$\begin{array}{r} 1,5\overset{4}{\,}\overset{1}{2}6 \\ -\ \ 873 \\ \hline 653 \\ \sim\sim\sim\sim\sim \\ 1,526 \end{array}$$
check:
$$\begin{array}{r} 873 \\ +\ 653 \\ \hline 1,526 \end{array}$$

2.
$$\begin{array}{r} \overset{3}{4},\overset{9}{0}\overset{9}{0}\overset{1}{0} \\ -\ 2,256 \\ \hline 1,744 \\ \sim\sim\sim\sim\sim \\ 4,000 \end{array}$$
check:
$$\begin{array}{r} 2,256 \\ +1,744 \\ \hline 4,000 \end{array}$$

3.
$$\begin{array}{r} 5,333 \\ -1,186 \\ \hline \end{array}$$

4.
$$\begin{array}{r} 4,284 \\ -\ \ 955 \\ \hline \end{array}$$

5.
$$\begin{array}{r} 8,263 \\ -3,149 \\ \hline \end{array}$$

6.
$$\begin{array}{r} 3,264 \\ -2,582 \\ \hline \end{array}$$

7.
$$\begin{array}{r} 7,241 \\ -\ \ 378 \\ \hline \end{array}$$

8.
$$\begin{array}{r} 6,000 \\ -5,139 \\ \hline \end{array}$$

9.
$$\begin{array}{r} 6,732 \\ -3,152 \\ \hline \end{array}$$

If you have trouble keeping numbers in the right places, you can use notebook paper turned sideways.

10. Micah counted 1,465 ants. He watched 906 of them go back into their hole. How many ants are left for Micah to watch?

   _____

11. It is 2,375 miles to Grandmother's house. Anne has already traveled 1,490 miles. How far does she still have to travel to get to Grandmother's house?

   _____

12. While on vacation, Kaylee saw one tree that was 1,760 years old and another tree that was 1,588 years old. What was the difference in their ages?

   _____

26B

Subtract using regrouping. Check by adding.

1.
$$
\begin{array}{r}
5,0\,8\,9 \\
-\ \ \ 6\,3\,2 \\
\hline
\end{array}
$$

2.
$$
\begin{array}{r}
7,3\,2\,1 \\
-\ 2,5\,1\,4 \\
\hline
\end{array}
$$

3.
$$
\begin{array}{r}
9,0\,0\,0 \\
-\ 1,2\,8\,7 \\
\hline
\end{array}
$$

4.
$$
\begin{array}{r}
7,1\,1\,1 \\
-\ \ \ 2\,3\,2 \\
\hline
\end{array}
$$

5.
$$
\begin{array}{r}
5,3\,6\,1 \\
-\ 3,7\,6\,5 \\
\hline
\end{array}
$$

6.
$$
\begin{array}{r}
7,2\,1\,4 \\
-\ 1,1\,0\,8 \\
\hline
\end{array}
$$

7.
$$
\begin{array}{r}
6,4\,0\,3 \\
-\ \ \ 2\,5\,7 \\
\hline
\end{array}
$$

8.
$$
\begin{array}{r}
8,7\,6\,5 \\
-\ 3,0\,8\,5 \\
\hline
\end{array}
$$

9.
$$
\begin{array}{r}
4,9\,8\,7 \\
-\ 3,7\,3\,2 \\
\hline
\end{array}
$$

10. Mom saw a set of living room furniture that cost $1,579. She has $890 saved. How much more money does she need to buy the set?

_____

11. Jeremy and Jeffrey lived far apart and wanted to visit each other. Jeremy traveled 3,451 miles, and Jeffrey traveled 2,999 miles. How many more miles did Jeremy travel?

_____

12. Joanne was born in 1976, and her grandfather was born in 1917. How old was her grandfather when Joanne was born? (Dates are a way of counting years. Subtract to find the number of years between two dates.)

_____

Subtract using regrouping. Check by adding.

1.
```
   1,6 5 0
 -   9 4 3
```

2.
```
   8,2 0 0
 - 2,8 1 7
```

3.
```
   5,2 2 1
 - 4,7 4 0
```

4.
```
   8,7 6 5
 -   6 7 8
```

5.
```
   2,9 0 6
 - 1,0 8 8
```

6.
```
   6,4 2 9
 - 3,5 8 7
```

7.
```
   3,4 0 5
 -   1 5 9
```

8.
```
   7,0 0 1
 - 5,9 9 1
```

9.
```
   9,3 2 4
 - 8,3 5 8
```

10. Kate learned that Columbus sailed across the ocean in 1492. Kate was born in 2004. How many years later was that?

   _____

11. Kimberly had $3,600 in her savings account. If she took out $150, how much was left in her savings?

   _____

12. A farmer harvested 2,562 bushels of corn. He sold 1,600 bushels. How many bushels of corn does he have left?

   _____

Subtract using regrouping as needed. Check by adding.

1.
$$\begin{array}{r} 4,305 \\ -289 \\ \hline \end{array}$$

2.
$$\begin{array}{r} 5,840 \\ -3,914 \\ \hline \end{array}$$

3.
$$\begin{array}{r} 7,013 \\ -1,523 \\ \hline \end{array}$$

4.
$$\begin{array}{r} 200 \\ -34 \\ \hline \end{array}$$

5.
$$\begin{array}{r} 772 \\ -416 \\ \hline \end{array}$$

6.
$$\begin{array}{r} 29 \\ -15 \\ \hline \end{array}$$

Show the number with tally marks.

7. 8 _____

8. 31 _____

9. 14 _____

Give the time with hours and minutes.

10.

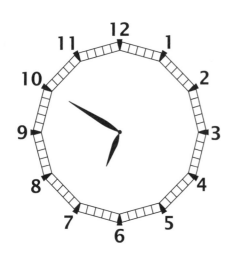

_____

11.

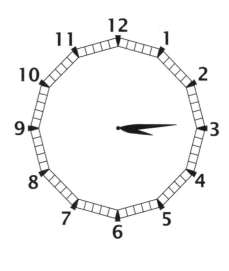

_____

12. _____ is the third day of the week.

13. December is the _____ month of the year.

14. Kara is five feet tall. How many inches tall is she?

_____

15. Stephanie traveled 451 miles the first day and 385 miles the second day. How many miles must she still travel to finish her 1,000-mile trip?

_____

26E

Subtract using regrouping as needed. Check by adding.

1.
```
  1,8 0 0
  -   1 7 6
```

2.
```
  6,9 5 0
  - 4,8 6 7
```

3.
```
  2,0 9 3
  - 2,0 7 5
```

4.
```
  7 4 1
  -  5 3
```

5.
```
  5 0 0
  - 2 1 1
```

6.
```
  4 8
  - 1 9
```

Show the number with tally marks.

7. 5 _____

8. 24 _____

9. 12 _____

Give the time with hours and minutes.

10.

_____

11.

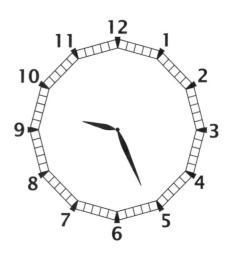

_____

12. _____ is the tenth month of the year.

13. Saturday is the _____ day of the week.

14. Alison collected 673 acorns. Chucky collected 415 acorns and added them to the pile. A few days later they could find only 223 acorns. How many acorns do you think the squirrels took away?

_____

15. What is the perimeter of a square that measures 10 inches on each side?

_____

Subtract using regrouping as needed. Check by adding.

1.
$$\begin{array}{r} 8,901 \\ -792 \\ \hline \end{array}$$

2.
$$\begin{array}{r} 9,012 \\ -8,999 \\ \hline \end{array}$$

3.
$$\begin{array}{r} 1,234 \\ -1,045 \\ \hline \end{array}$$

4.
$$\begin{array}{r} 670 \\ -69 \\ \hline \end{array}$$

5.
$$\begin{array}{r} 765 \\ -578 \\ \hline \end{array}$$

6.
$$\begin{array}{r} 32 \\ -25 \\ \hline \end{array}$$

Show the number with tally marks.

7. 11   _____

8. 33   _____

9. 26   _____

Give the time with hours and minutes.

10.

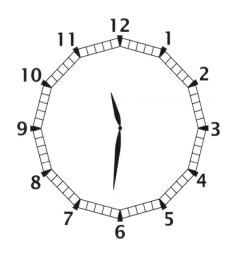

_____

11.

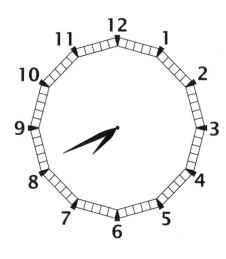

_____

12. _____ is the fifth day of the week.

13. June is the _____ month of the year.

14. Christel called two friends on Monday, four on Tuesday, five on Wednesday, six on Thursday, and eight on Friday. How many phone calls did she make on those five days?

_____

15. What is the perimeter of a rectangle that is six inches long and three inches high?

_____

Subtract.

If the answer is 50, color the space brown.
If the answer is 75, color the space blue.
If the answer is 124, color the space green.
If the answer is 267, leave the space white.

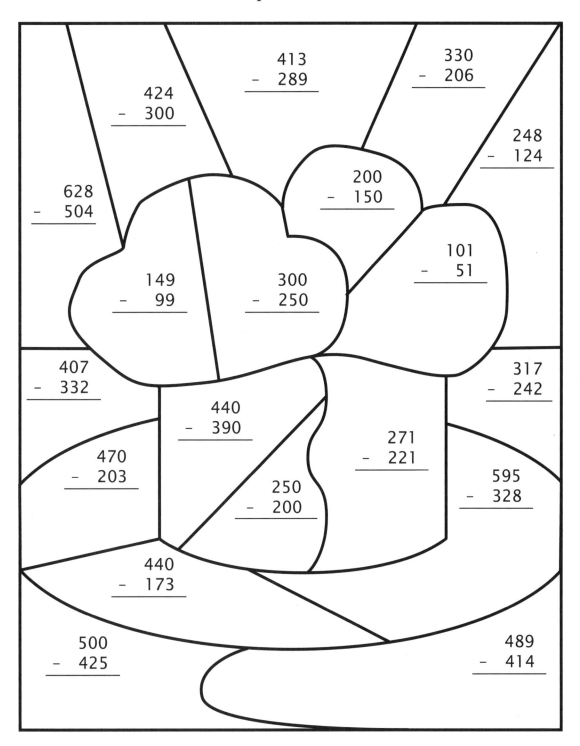

Help to write the word problems. Fill in names of people you know. Write your answers in the boxes.

1. _____ counted 1,467 stars on Monday night. On Tuesday night 2,345 stars were counted. How many more stars were counted on Tuesday night?

2. _____ captured 1,681 dandelion seeds and put them in a bag. A strong gust of wind blew 592 of seeds away. How many dandelion seeds are left in the bag?

For this question, fill in names of two people you know, and the years they were born. Ask your teacher if you need help.

3. _____ was born in _____.

_____ was born in _____.

Subtract the years to find the differences in their ages.

Subtract the amounts of money. The first one has been done for you.

1.
$$\begin{array}{r} {\scriptstyle 2\ 1} \\ \$1.\cancel{3}4 \\ -\ 0.75 \\ \hline \$0.59 \end{array}$$

2.
$$\begin{array}{r} \$9.00 \\ -\ 2.67 \\ \hline \end{array}$$

3.
$$\begin{array}{r} \$0.46 \\ -\ 0.10 \\ \hline \end{array}$$

4.
$$\begin{array}{r} \$6.14 \\ -\ 1.21 \\ \hline \end{array}$$

5.
$$\begin{array}{r} \$2.45 \\ -\ 1.38 \\ \hline \end{array}$$

6.
$$\begin{array}{r} \$0.80 \\ -\ 0.24 \\ \hline \end{array}$$

7.
$$\begin{array}{r} \$5.19 \\ -\ 3.72 \\ \hline \end{array}$$

8.
$$\begin{array}{r} \$26.00 \\ -\ 11.92 \\ \hline \end{array}$$

9.
$$\begin{array}{r} \$34.47 \\ -\ 22.09 \\ \hline \end{array}$$

10. Ethan had $7.15 when he went into the store and $2.98 when he came out. How much did he spend?

    _____

11. Bria had $9.46 in her purse. She decided to spend $4.91 on a gift for Duncan. How much did she have left?

    _____

12. My boots cost $45.50, and my shoes cost $34.99. How much more did my boots cost?

    _____

Subtract the amounts of money.

1. $2.6 5
    − 0.3 8

2. $7.2 5
    − 1.8 9

3. $0.1 0
    − 0.0 8

4. $1.0 0
    − 0.7 7

5. $6.1 9
    − 3.5 8

6. $0.9 3
    − 0.4 5

7. $2.4 7
    − 1.0 9

8. $3 5.6 0
    − 2 1.9 0

9. $7 4.8 2
    − 3 6.2 5

10. Matthew and Josh bought CDs. Matthew spent $25.69, and Josh spent $31.16. How much more did Josh spend?

   _____

11. The game that Abby wants to buy costs $39.67. She has saved $20.75. How much more does she need to save?

   _____

12. Caitlyn earned $13.05 helping her mom. After she went shopping, she had only $0.36 left. How much money did Caitlyn spend?

   _____

Subtract the money.

1. $5.1 3
     − 0.4 8

2. $8.1 5
     − 7.0 9

3. $ 0.7 3
     − 0.1 7

4. $3.0 0
     − 0.8 1

5. $9.3 2
     − 6.1 4

6. $ 0.8 6
     − 0.5 5

7. $4.7 5
     − 2.0 6

8. $2 0.1 5
     − 1 3.2 1

9. $9 8.1 7
     − 2 5.1 8

10. Sue took $10.00 to the store to buy fabric for a quilt. She came home with $1.25. How much did she spend?

_____

11. Terri wants to buy a gift for Mindy that costs $53.95. She has saved $45.00 already. How much more does she need?

_____

12. Bananas cost 45¢ a pound, and apples cost 61¢ a pound. How much more would it cost to buy a pound of apples?

_____

Subtract.

1. $8.91
    − 0.76

2. $6.82
    − 1.61

3. 7,712
    − 5,872

Add.

4. 184
    + 68

5. 255
    + 177

6. 3,011
    + 1,895

Matching.

7. first   Wednesday

8. second   Friday

9. third   Saturday

10. fourth   Sunday

11. fifth   Thursday

12. sixth   Monday

13. seventh   Tuesday

Skip count by two.

14. 2, ____, ____, ____, ____, ____, ____, ____, ____, ____

15. Rick kept track of the birds that came to his feeder one morning: ‖‖‖ ‖‖‖ ‖‖‖ ‖. How many birds did he see?

  _____

16. Chuck got $5.00 from each of his four sisters for his birthday. Use column addition to find out how much he got in all. Chuck spent $6.95 for a ball. How much money did he have left?

  _____

Subtract.

1. $4.4 2
   − 0.3 9

2. $1.2 0
   − 1.1 8

3.   4,5 0 3
   − 2,9 0 1

Add.

4.   8 3 6
   + 1 7

5. $3.4 5
   + 7.1 8

6.   1,8 1 9
   + 4,4 2 8

Matching.

7. first   June

8. second   April

9. third   January

10. fourth   March

11. fifth   February

12. sixth   May

Skip count by five.

13. 5, ____, ____, ____, ____, ____, ____, ____, ____, ____

14. Hannah kept track of the days until the package she had ordered came in the mail: ||||| ||||| ||||| ||||| ||||| |||||. How many days did she wait?

_____

15. Drew did two chores. He earned $5.00 and $6.50. Cameron did one chore. He earned $10.20. Find out which boy earned more. Can you remember how to use an inequality to show your answer?

_____

Subtract.

1. $3.14
   − 1.09

2. $2.62
   − 1.38

3. $19.07
   − 15.21

Add.

4. 977
   + 13

5. $6.54
   + 9.54

6. 1,764
   1,913
   + 2,384

Matching.

7. seventh   August

8. eighth   October

9. ninth   December

10. tenth   July

11. eleventh   November

12. twelfth   September

Skip count by ten.

13. 10, _____, _____, _____, _____, _____, _____, _____, _____, _____

14. Use tally marks to show the number of days in August.

      _____

15. A fence went all around the perimeter of a triangle that had sides of 35 feet, 25 feet, and 19 feet. Ten feet of fence got knocked down when Bill hit it with his go-cart. How many feet of fence are left standing?

      _____

Subtract to find the change that each person should get back. Draw a line around the box that shows the correct change.

1. A book costs $4.79. Sophia gave the cashier $5.00. What should the cashier give back to her?

2. Daniel bought a sandwich that cost $2.64. He gave the server $5.00. What should the server give back to him?

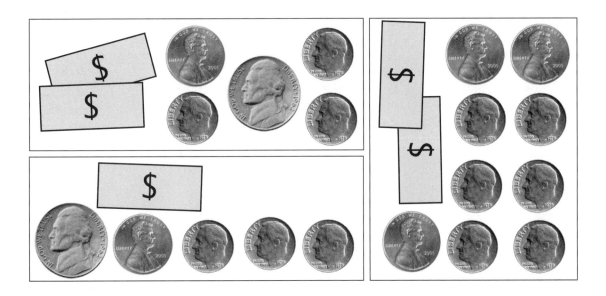

3. A game costs $8.99. Madison gave the cashier $10.00. How much money should the cashier give back to her?

4. A box of crayons costs $2.85. Sara gave the cashier $5.00. How much money should the cashier give back to her?

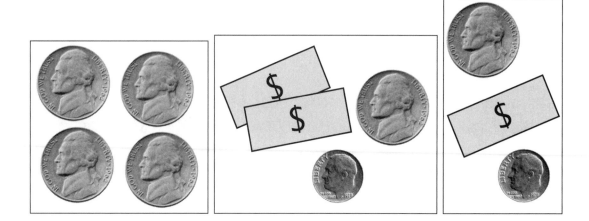

Subtract using regrouping. Check your work. The first one has been done for you.

1.
$$\begin{array}{r} {}^1\overset{}{2}{}^1\!1,{}^6\overset{}{7}\,{}^{15}\overset{}{6}{}^1\!2 \\ -\ 1\ 4,\ 3\ 7\ 5 \\ \hline 7,\ 3\ 8\ 7 \\ \hline 2\ 1,\ 7\ 6\ 2 \end{array}$$

check:
$$\begin{array}{r} 1\ 4,\ 3\ 7\ 5 \\ +\ \ \ 7,\ 3\ 8\ 7 \\ \hline 2\ 1,\ 7\ 6\ 2 \end{array}$$

2.
$$\begin{array}{r} 8\ 4,5\ 2\ 8 \\ -\ 6\ 4,0\ 2\ 5 \\ \hline \end{array}$$

3.
$$\begin{array}{r} 3\ 5,1\ 9\ 4 \\ -\ 3\ 1,4\ 8\ 6 \\ \hline \end{array}$$

4.
$$\begin{array}{r} 4\ 2,3\ 5\ 5 \\ -\ 2\ 1,4\ 7\ 2 \\ \hline \end{array}$$

5.
$$\begin{array}{r} 7\ 1,6\ 2\ 1 \\ -\ 4\ 1,5\ 7\ 3 \\ \hline \end{array}$$

6.
$$\begin{array}{r} 5\ 6,4\ 0\ 8 \\ -\ 2\ 4,3\ 7\ 9 \\ \hline \end{array}$$

7. Our town used to have 45,900 people. Then 22,175 people moved away. How many people live in the town now?

_____

8. This spring, 31,231 tadpoles hatched out of their eggs. Before they could grow up to be frogs, 19,452 of them were eaten. How many tadpoles were left to become frogs?

_____

9. The distance around the earth is about 25,000 miles. Thomas set out to fly all the way around in his plane. If he has flown 19,000 miles so far, how many miles does he still have to go?

_____

10. Ruth thought she would earn $45,575 last year. Since she was sick a lot and couldn't go to work, she earned only $38,196. What is the difference between the two amounts?

_____

Subtract using regrouping. Check your work.

1.
$$
\begin{array}{r}
6\,8{,}5\,1\,1 \\
-\ 1\,9{,}3\,3\,3 \\
\hline
\end{array}
$$

2.
$$
\begin{array}{r}
2\,5{,}3\,8\,2 \\
-\ 1\,1{,}2\,5\,5 \\
\hline
\end{array}
$$

3.
$$
\begin{array}{r}
4\,7{,}4\,6\,0 \\
-\ 3\,3{,}4\,1\,9 \\
\hline
\end{array}
$$

4.
$$
\begin{array}{r}
7\,0{,}0\,0\,0 \\
-\ 1\,9{,}9\,9\,9 \\
\hline
\end{array}
$$

5.
$$
\begin{array}{r}
6\,5{,}2\,4\,2 \\
-\ 2\,1{,}1\,3\,5 \\
\hline
\end{array}
$$

6.
$$
\begin{array}{r}
5\,4{,}9\,6\,7 \\
-\ 4\,2{,}7\,1\,8 \\
\hline
\end{array}
$$

7. In the town election, 10,752 people voted. Of that number, 6,834 people voted for Mr. Jones for mayor. How many people voted for someone else?

_____

8. A farmer planted 55,780 seeds of corn. Because it was so dry, 29,592 seeds did not sprout. How many corn plants did start to grow?

_____

9. Rebecca figured out that her one-mile walk was 63,360 inches long. Clara walked only 31,680 inches. What is the difference in the number of inches they walked?

_____

10. Ben is thinking of buying a new car. One car he likes costs $16,899, and another costs $22,980. What is the difference in price between the two cars?

_____

Subtract using regrouping. Check your work.

1.
```
  1 3,5 2 2
- 1 2,0 4 8
```

2.
```
  7 9,1 4 7
- 2 4,3 1 2
```

3.
```
  5 3,6 4 1
- 2 0,4 6 5
```

4.
```
  8 4,6 7 0
- 3 5,9 0 1
```

5.
```
  1 3,5 0 6
-   9,9 5 1
```

6.
```
  4 1,0 0 3
- 1 0,5 2 3
```

7. The odometer in a car measures how many miles the car has been driven. When Elizabeth left for vacation, the odometer in her car read 19,761 miles. When she got home, it read 23,295 miles. How many miles had she driven?

   _____

8. The plastics company made 53,600 green unit blocks for Math-U-See. If 48,096 have been packed in the boxes to sell, how many unit blocks are left to pack?

   _____

9. A scientist estimated that 60,000 fish lived in a lake. He thought that about 22,000 of the fish were trout. If he is correct, about how many fish are not trout?

   _____

10. Katherine wants to buy a house that costs $95,899. She has $26,900. How much more money does she need?

    _____

Subtract using regrouping as needed. Check your work.

1.
$$\begin{array}{r} 9\,6{,}0\,2\,1 \\ -\ 4\,5{,}6\,3\,5 \\ \hline \end{array}$$

2.
$$\begin{array}{r} 8\,0{,}0\,0\,0 \\ -\ 7\,9{,}9\,9\,8 \\ \hline \end{array}$$

3.
$$\begin{array}{r} \$5\,0.7\,5 \\ -\ 1\,0.2\,5 \\ \hline \end{array}$$

4.
$$\begin{array}{r} \$3\,9.1\,3 \\ -\ 2\,2.4\,6 \\ \hline \end{array}$$

5.
$$\begin{array}{r} \$1\,9.1\,5 \\ -\ 7.9\,9 \\ \hline \end{array}$$

Fill in the oval with >, <, or =.

6. 3 $\bigcirc$ 4

7. 2 + 5 $\bigcirc$ 2 + 4

8. 3 + 6 $\bigcirc$ 6 + 3

Add.

9.
$$
\begin{array}{r}
1\,2\,5 \\
2\,8\,5 \\
4\,4 \\
+\ 1\,6\,1 \\
\hline
\end{array}
$$

10.
$$
\begin{array}{r}
3\,1\,5 \\
2\,0\,0 \\
3\,9\,4 \\
+\ 1\,5\,5 \\
\hline
\end{array}
$$

11.
$$
\begin{array}{r}
4\,4\,9 \\
1\,3\,1 \\
5\,0\,3 \\
+\ \ 6\,7 \\
\hline
\end{array}
$$

12. _____ is the first day of the week.

13. March is the _____ month of the year.

14. In April, Mom went grocery shopping four times. She spent $123.45, $210.13, $75.21, and $103.82. How much did Mom spend for food during that month?

_____

15. A shepherd had 100 sheep, but when he counted them one night he found only 99 sheep. How many sheep were lost?

_____

Subtract using regrouping as needed. Check your work.

1.
$$\begin{array}{r} 1\,1,4\,3\,5 \\ -\ 1\,0,6\,8\,2 \\ \hline \end{array}$$

2.
$$\begin{array}{r} 4\,0,7\,3\,4 \\ -\ 2\,7,1\,5\,6 \\ \hline \end{array}$$

3.
$$\begin{array}{r} \$2\,1.1\,1 \\ -\ 1\,9.8\,9 \\ \hline \end{array}$$

4.
$$\begin{array}{r} \$7\,8.0\,4 \\ -\ 3\,5.4\,0 \\ \hline \end{array}$$

5.
$$\begin{array}{r} \$6\,2.0\,0 \\ -\ 5\,1.1\,9 \\ \hline \end{array}$$

Fill in the oval with >, <, or =.

6. 14 $\bigcirc$ 7 + 7

7. 9 – 7 $\bigcirc$ 7 – 4

8. 63 $\bigcirc$ 36

Add.

9.
```
  7 0 4
  3 0 0
    2 6
+     9
```

10.
```
    2 4 1
    1 1 9
    5 6 2
+ 5 0 8
```

11.
```
  3 2 5
  7 6 3
  1 2 5
+   4 6
```

12. _____ is the eighth month of the year.

13. Friday is the _____ day of the week.

14. Last week, 10,456 spiderlings hatched in my garden. The birds ate 8,912 of them. How many spiderlings grew up to be big spiders?

_____

15. Farmer Brown has 25 cows, 89 sheep, and 5 horses. How many animals does he have in all?

_____

Subtract using regrouping as needed. Check your work.

1.
$$
\begin{array}{r}
7\,9,3\,0\,4 \\
-\ 6\,1,0\,8\,2 \\
\hline
\end{array}
$$

2.
$$
\begin{array}{r}
5\,5,0\,0\,0 \\
-\ 4\,8,1\,2\,3 \\
\hline
\end{array}
$$

3.
$$
\begin{array}{r}
\$1\,7.9\,9 \\
-\ 5.1\,5 \\
\hline
\end{array}
$$

4.
$$
\begin{array}{r}
\$4\,3.6\,0 \\
-\ 1\,3.7\,9 \\
\hline
\end{array}
$$

5.
$$
\begin{array}{r}
\$2\,8.7\,4 \\
-\ 1\,3.2\,5 \\
\hline
\end{array}
$$

Fill in the oval with >, <, or =.

6. 42 $\bigcirc$ 78

7. 14 + 2 $\bigcirc$ 13 − 2

8. 2 + 1 $\bigcirc$ 1 + 2

Add.

9.
```
  9 1 0
  2 6 2
  3 6 6
+ 1 4 8
```

10.
```
  3 5 5
  1 7 6
  7 5 5
+   2 3
```

11.
```
  4 3 1
  5 2 9
  6 1 1
+ 5 7 6
```

12. _____ is the fourth day of the week.

13. May is the _____ month of the year.

14. Caleb got $34.00 for his birthday and $25.00 for Christmas. He spent $29.98 on a gift for his sister. How much money does he have left?

    _____

15. Joel counted cars during his ride. He counted 25 red cars, 17 black cars, 30 blue cars, and 8 white cars. How many cars did he count altogether?

    _____

You can use your column addition skills to find objects arranged in a rectangular group called an *array*.

```
      3
      3
      3
   +  3
     12
```

3 + 3 + 3 + 3 = 12 flowers

Use column addition to find the number of objects in each array.

1.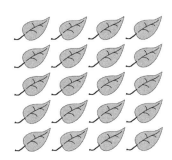

4 + 4 + 4 + 4 + 4 = _____ leaves

2.

_ + _ + _ + _ + _ = __ butterflies

3.

_ + _ + _ + _ + _ = __ stars

4.

_ + _ + _ = __ snowflakes

Are the math words you learned this year all mixed up? Here is a chance to unscramble some of them. Use the unscrambled words at the bottom of the page for clues.

msu _____

geiantrl _____

naddde _____

lpaec lveau _____ _____

usaqre _____

remetipre _____

ctraeglne _____

cfeedfrnie _____

yllta smrka _____ _____

---

addend	place value	sum
difference	rectangle	tally marks
perimeter	square	triangle

Fill in the missing numbers. Write the number of gallons on the line. The first one has been done for you.

1.

___15___ gallons

2.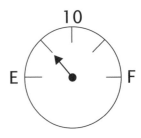

_____ gallons

3.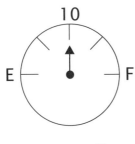

_____ gallons

What speed does each speedometer show? (The letters mph stand for miles per hour.)

4.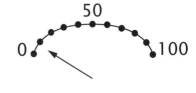

_____ mph

5.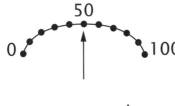

_____ mph

6.

_____ mph

Fill in the missing numbers. Write the temperature on the line below. The first one has been done for you.

7.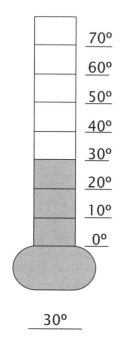

70°
60°
50°
40°
30°
20°
10°
0°

__30°__

8.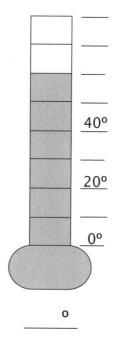

40°

20°

0°

___ °

9.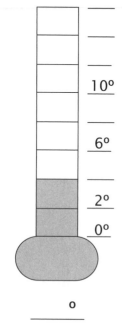

10°

6°

2°
0°

___ °

Fill in the missing numbers on the gauge. Write the temperature on the line below. The first one has been done for you.

1.
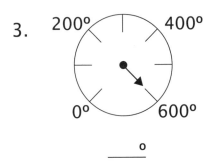

__400°__

2.
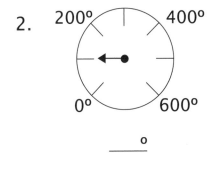

____°

3.
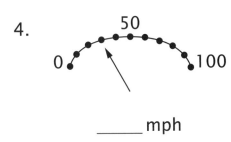

____°

What speed does each speedometer show?

4.
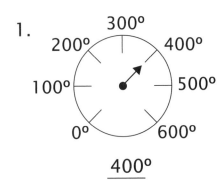

_____ mph

5.
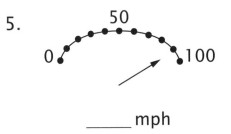

_____ mph

6.
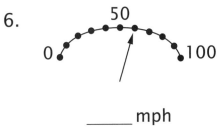

_____ mph

Fill in the missing numbers. Write the temperature on the line below.

7.

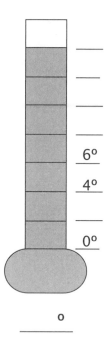

___
___
___
6°
4°
___
0°

o

_____

8.

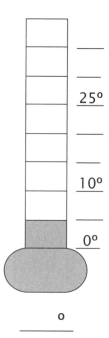

___
___
25°
___
___
10°
___
0°

o

_____

9.

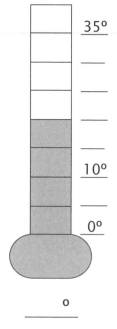

35°
___
___
___
___
10°
___
0°

o

_____

29C

Fill in the missing numbers. Write the number of gallons on the line.

1.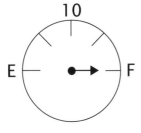

_____ gallons

2.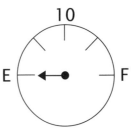

_____ gallons

3.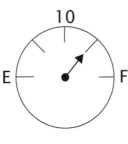

_____ gallons

What speed does each speedometer show?

4.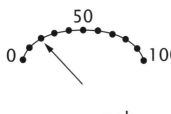

_____ mph

5.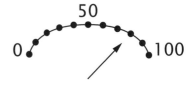

_____ mph

6.

_____ mph

Fill in the missing numbers. Write the temperature on the line below.

7.

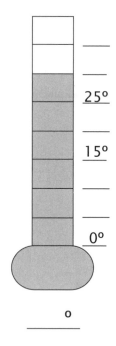

_____
_____
_25°_
_15°_
_____
_____
_0°_

o

_____

8.

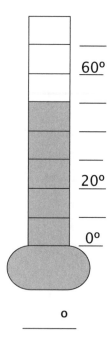

_60°_
_____
_____
_____
_20°_
_____
_0°_

o

_____

9.

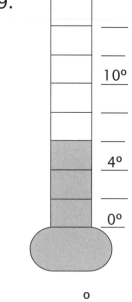

_____
_10°_
_____
_4°_
_____
_0°_

o

_____

29D

Read the gauge or thermometer and write your answer on the line.

1.

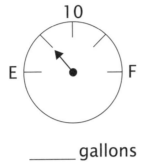

_____ gallons

2.

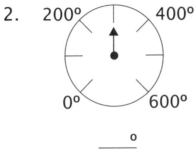

_____ °

3.

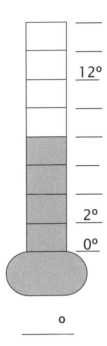

_____ °

4.

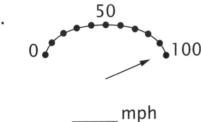

_____ mph

5.

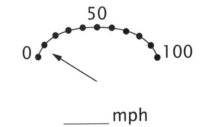

_____ mph

Add or subtract.

6.
$$
\begin{array}{r}
3\,4{,}1\,2\,7 \\
-\ 2\,5{,}1\,7\,9 \\
\hline
\end{array}
$$

7.
$$
\begin{array}{r}
5{,}6\,8\,0 \\
-\ 4{,}4\,5\,2 \\
\hline
\end{array}
$$

8.

$$
\begin{array}{r}
7,5\,6\,3 \\
+\ 5,1\,4\,8 \\
\hline
\end{array}
$$

Give the time with hours and minutes.

9.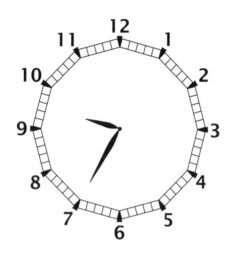

_____

10.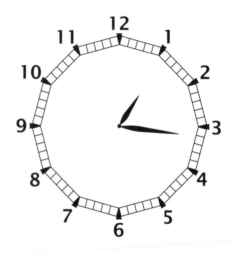

_____

11. Peter was born in 1962. How many years old was Peter in the year 2003?

_____

12. Use tally marks to show Peter's age in 2003 (see #11).

_____

Read the gauge or thermometer and write your answer on the line.

1.

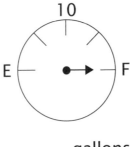

_____ gallons

2.

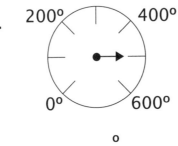

_____ °

3.

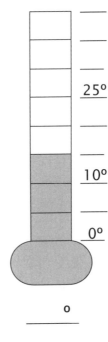

_____ °

4.

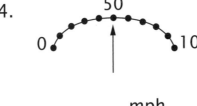

_____ mph

5.

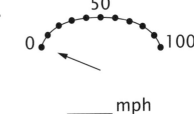

_____ mph

Add or subtract.

6.
$$\begin{array}{r} 17,539 \\ -\ 15,659 \\ \hline \end{array}$$

7.
$$\begin{array}{r} 7,058 \\ -\ 3,172 \\ \hline \end{array}$$

8.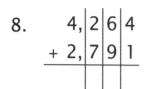

QUICK TIP

Sequencing can be used to find your room in a hotel. The first digit in the room number tells the floor. Once you are on the correct floor, the rooms are numbered in order.

Example 1
When you get off the elevator on the first floor, you see this sign.

Do you turn to the right or the left to find room 172?

101 – 145	146 – 190
←	→

Answer: You would turn right because 172 comes between 146 and 190.

9. Using the sign above, tell if you should turn right or left to find room 116. _____

10. Our country was 200 years old on July 4, 1976. In what year did our country begin? _____

July is the _____ month.

Read the gauge or thermometer and write your answer on the line.

1.

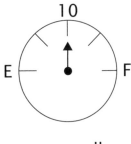

_____ gallons

2.

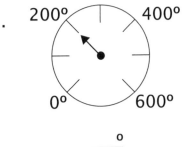

_____ °

3.

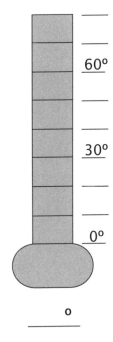

_____ °

4.

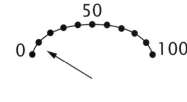

_____ mph

5.

_____ mph

Add or subtract.

6.
$$\begin{array}{r} 9\,3,1\,0\,5 \\ -\ 6\,0,1\,2\,7 \\ \hline \end{array}$$

7.
$$\begin{array}{r} 8,0\,0\,0 \\ -\ 2,1\,1\,1 \\ \hline \end{array}$$

8.
```
    6,4 5 6
    2,4 3 4
  + 1,8 4 9
```

Give the time with hours and minutes.

9.

_____

10.

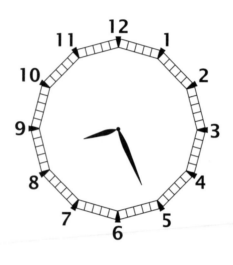

_____

11. Using the sign, tell if you should turn right or left to find room 430.

401 – 450	451 – 490
←	→

_____

12. The United States was 200 years old in 1976. How old is it now? (two steps)

_____

Use the thermometers on this page and the next one to record the daily temperatures for six days. Before you begin, help your child fill in the temperature scale on the lines. We suggest counting by fives and choosing a range of temperatures that will fit your season and climate. Write the day or date under each thermometer. Try to record the temperature at about the same time every day.

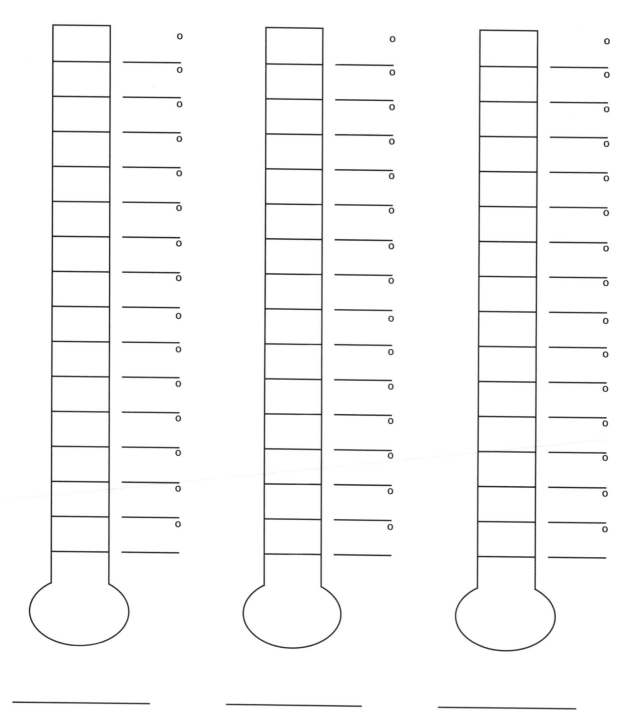

Which day was the coldest? _____

What was the temperature on that day?_____

Which day was the warmest? _____

What was the temperature on that day?_____

Use the bar graph to answer questions 1-4.

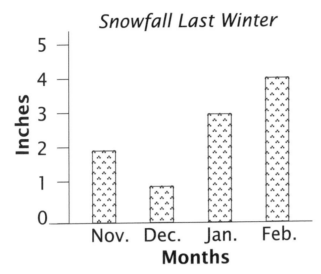

Snowfall Last Winter

1. Which month had the most snow? _____

2. Which month had the least snow? _____

3. How much snow fell in November? _____

4. Which month had 3 inches of snow? _____

5. The year before, 3" fell in November, 2" fell in December, 4" fell in January, and 1" fell in February. Make a line graph to show the change.

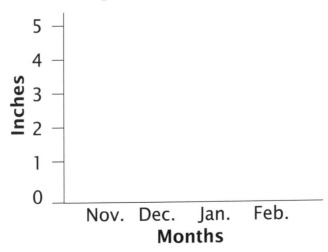

Change in Snowfall Last Winter

A bar graph can be turned on its side. This is good for showing length or distance.

Shoe Sizes of Children in the Family

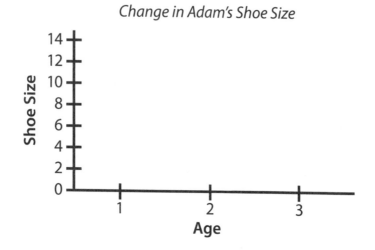

6. Who has the biggest shoe size? _____

7. What is the smallest shoe size? _____

8. What is David's shoe size? _____

9. Which two children have the same shoe size?

    _____

10. When Adam was 1 year old, he wore a size 3 shoe. When he was 2, he wore a size 6. Now he is 3 and wears a size 8. Make a line graph to show how Adam's shoe size has changed.

Change in Adam's Shoe Size

BETA

Use the bar graph to answer questions 1-4.

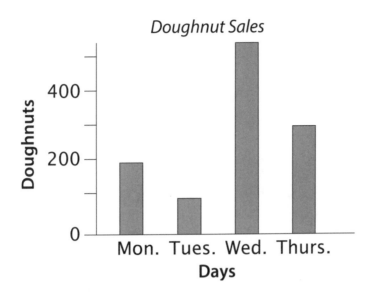

1. Which day had the most sales? _____

2. Which day had the fewest sales? _____

3. How many doughnuts were sold on Monday? _____

4. How many doughnuts were sold on Thursday? _____

5. The next week's sales were 200, 300, 100, and 400. Make a line graph to show the change from day to day.

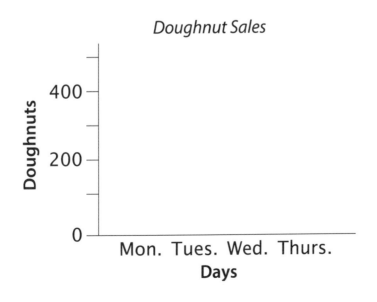

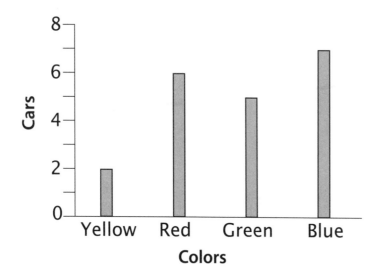

Colors of Cars Passing Bill's House

6. Which color did he see the most? _____

7. How many red cars did he see? _____

8. How many green cars did he see? _____

9. How many yellow cars did he see? _____

10. The next day he saw 1 yellow car, 4 red cars, 8 green cars, and 5 blue cars. Make a bar graph.

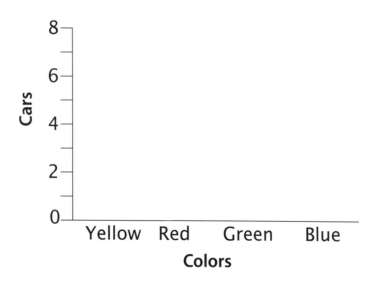

Colors of Cars Passing Bill's House

Use the bar graph to answer questions 1-3.

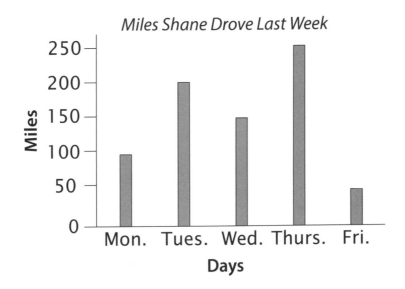

1. How many miles did he drive on Monday? _____

2. How many miles did he drive on Friday? _____

3. How many miles did Shane drive all week? _____

4. Shane wanted to graph his speed on Friday. After 5 miles, he was traveling at 30 mph. After 10 miles, he was traveling at 50 mph. Between 35 and 40 miles, he slowed down to 20 mph. Make a line graph to show this.

Shane's Speed on Friday

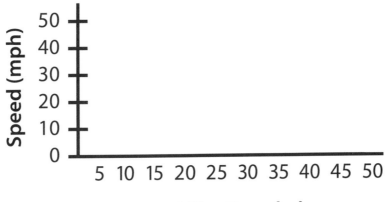

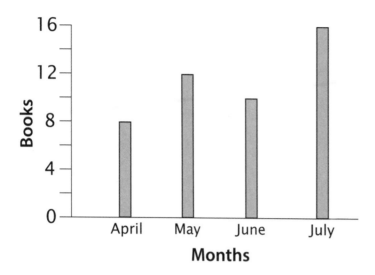

Books Sarah Read Each Month

5. How many books did Sarah read in July? _____

6. How many books did she read in May? _____

7. How many books did Sarah read in the four months? _____

8. The next four months Sarah read the following numbers of books each month: 10, 12, 8, and 12. Make a bar graph.

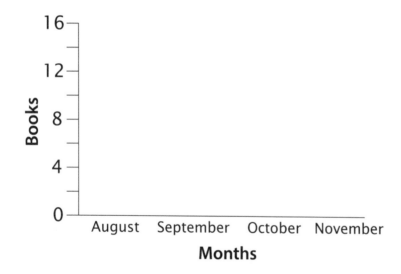

Number of Books Sarah Read

Use the bar graph to answer questions 1-3.

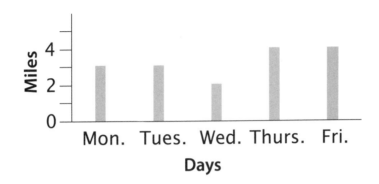

Miles Isaac Walked This Week

1. On which days did Isaac walk the farthest?

   _____ and _____

2. On which day did he walk the fewest miles? _____

3. How many miles did Isaac walk all week? _____

4. The same week Ethan walked the following numbers of miles each day: 3, 4, 1, 2, and 4. Make a line graph to show the change over the five days.

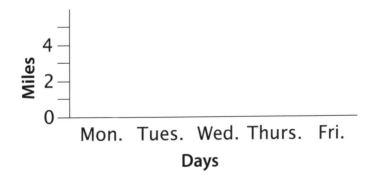

Change in Miles Ethan Walked

Read the gauge and write your answer on the line.

5.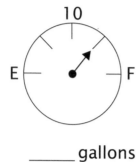

_____ gallons

6.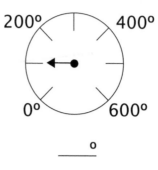

_____ °

Add or subtract.

7. $7 8.9 0
   – 2 1.9 5

8. $1 3.5 4
   + 8.6 3

9.   5 3,4 1 1
   – 1 6,4 2 2

10. Using the sign, tell if you should turn right or left to find room 761.

701 – 750	751 – 790
←	→

_____

11. Write the number shown by the tally marks: ⅢⅡ ⅢⅡ ⅢⅡ ⅢⅡ ⅢⅡ I.

_____

12. _____ is the third month.

Use the bar graph to answer questions 1-3.

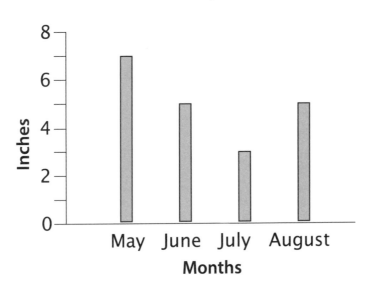

Monthly Rainfall

1. Which month had the most rain? _____

2. Which month had the least rain? _____

3. How much rain fell in August? _____

4. The next four months had the following amounts of rain: 4", 6", 8", and 5". Make a line graph.

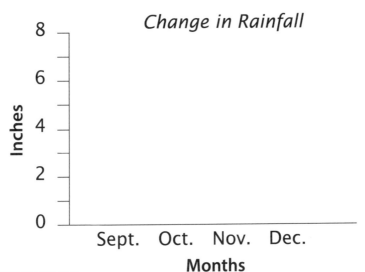

Change in Rainfall

Read the gauge and write your answer on the line.

5.

_____ mph

6.

_____ mph

Fill in the oval with >, <, or =.

7. 135 ◯ 531

8. 7 + 8 ◯ 5 + 5 + 5

9. 16 – 8 ◯ 5 + 4

10. Each side of a square measures 45 feet. What is the perimeter of the square?

_____

11. Jeff was born in 1945, and Jerry was born in 1972. What is the difference in their ages?

_____

12. Lisa drove 65 miles on Monday, 348 miles on Tuesday, and 179 miles on Wednesday. How many miles did she drive in those three days?

_____

Use the graph to answer the questions. Remember that bar graphs can be turned on their sides.

Home Runs in April

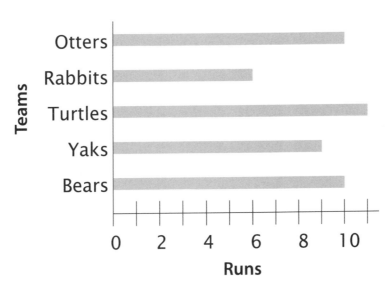

1. Which team hit the most home runs? _____

2. Which team hit the fewest home runs? _____

3. How many home runs did the Yaks hit? _____

4. Which two teams had the same number of home runs?

   _____ and _____

Read each thermometer and write your answer on the line below it.

5.

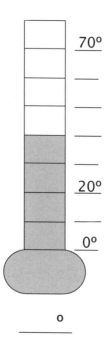

70°
—
—
—
20°
—
0°

o
___

6.

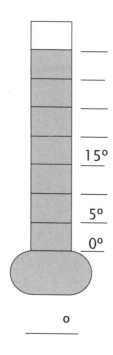

—
—
15°
—
5°
0°

o
___

7.

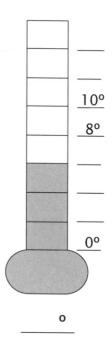

—
—
10°
8°
—
0°

o
___

Add or subtract.

8.
$$\begin{array}{r} 76,541 \\ -\ 32,153 \\ \hline \end{array}$$

9.
$$\begin{array}{r} 6,000 \\ -\ 3,265 \\ \hline \end{array}$$

10.
$$\begin{array}{r} 3,128 \\ 5,592 \\ +\ 4,517 \\ \hline \end{array}$$

Pictures may also be used to make a graph. The picture graph below shows how many of each kind of toy David has.

David's Toys

cars	trucks	baseballs
🏎		
🏎		⚾
🏎	🚚	⚾
🏎	🚚	⚾
🏎	🚚	⚾

Use the picture graph to answer the questions.

1. How many cars does David have?

2. How many trucks does David have?

3. How many more cars than trucks does he have?

4. How many toys have wheels?

5. How many toys do not have wheels?

6. How many more baseballs are needed to equal the number of toys with wheels?

Get 3 dimes, 4 nickels, and 2 pennies. Sort the coins by kind and then use the results to make a picture graph. Answer the questions.

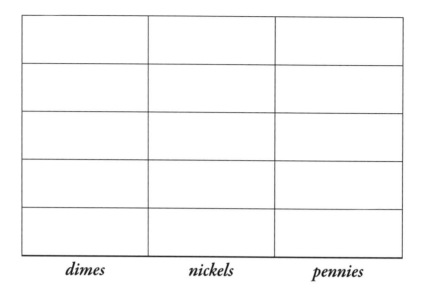

dimes   nickels   pennies

1. How many more dimes are needed to equal the number of nickels on the graph?

2. How many coins are silver-colored?

3. How many more pennies are needed to equal the number of silver-colored coins?

4. What is the total number of coins?

Sort objects and make your own graphs.

Another kind of graph is called a line plot. On a line plot, marks are made above a line to show the number of each kind. For example, Sue has five pencils. One pencil is 5 inches long, three pencils are each 3 inches long, and two pencils are each 2 inches long. The line plot looks like this.

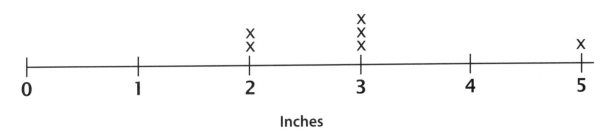

Lengths of Sue's Pencils

Inches

Use a ruler to measure the rectangles shown below. Use the line plot at the bottom of the page to show the length of each rectangle.

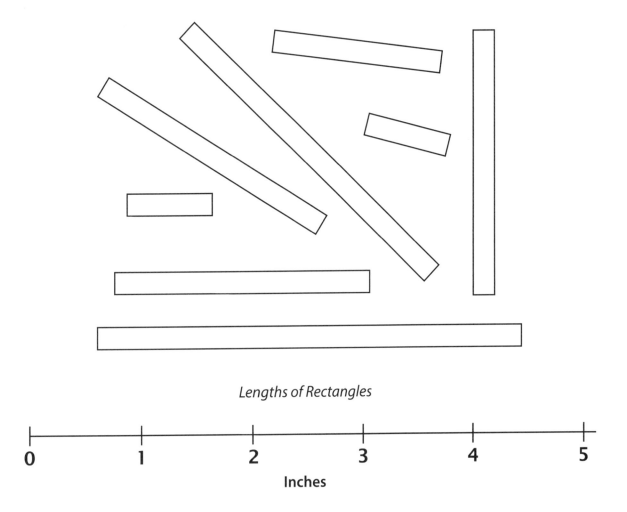

Lengths of Rectangles

Inches

Talking about equal shares is a good way to understand fractions. Help the student with these activities and use words such as halves, thirds, and fourths in everyday life.

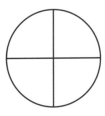

The cake is cut into 4 equal shares. Color a fourth of the cake to show one person's share.

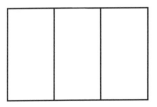

This cake is cut into 3 equal shares. Color a third of the cake to show one person's share.

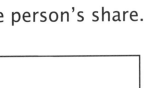

Color half of the rectangle.

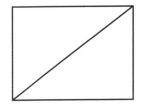

Color the halves of this rectangle two different colors.

Equal shares can have different shapes. This cake is cut into twelve pieces. If two people share it equally, each one would get six pieces. Color each cake to show the shares in two different ways.

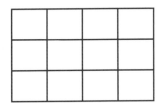

Color half of the cake.

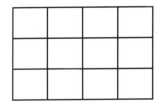

Color half of this cake, but make the share a different shape.

Shapes may be named by the number of their sides or by the number of their corners (angles). Students already know that triangles have three straight sides and three angles. Rectangles have four straight sides and four square corners. There are many four-sided shapes that are not rectangles. The general name that includes all four-sided shapes is quadrilateral. Here are some names that describe shapes.

A *triangle* has 3 straight sides and 3 angles.

A *quadrilateral* has 4 straight sides and 4 angles.

A *pentagon* has 5 straight sides and 5 angles.

A *hexagon* has 6 straight sides and 6 angles.

Follow the directions. Use another piece of paper if you wish.

1. Draw a shape with three angles and write its name.

2. Draw a shape with five straight sides and write its name.

3. Draw a shape with four angles and write its name.

4. Draw a shape with six straight sides and write its name.

5. Draw a shape with five angles and write its name.

B1

Addition and subtraction can be shown with the blocks. You may also show addition and subtraction with a number line. A number line looks something like a ruler. We can show six plus four with the blocks, on a number line, and with numerals.

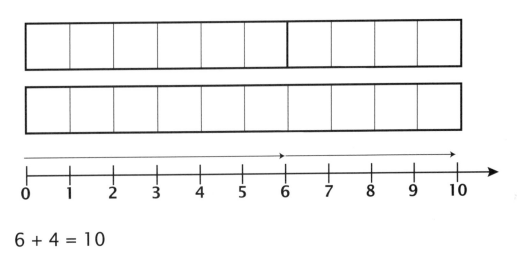

6 + 4 = 10

1. Use arrows to show five plus one on the number line.

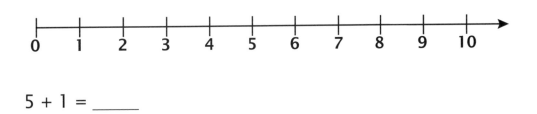

5 + 1 = _____

2. Use arrows to show four plus three on the number line.

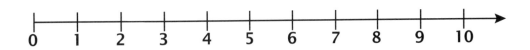

4 + 3 = _____

Here is an example of subtraction using blocks and a number line.

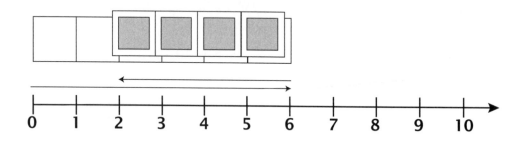

6 – 4 = 2

3. Use arrows to show eight minus three on the number line.

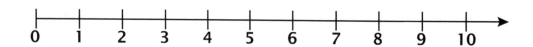

8 – 3 = _____

4. Use arrows to show four minus one on the number line.

4 – 1 = _____

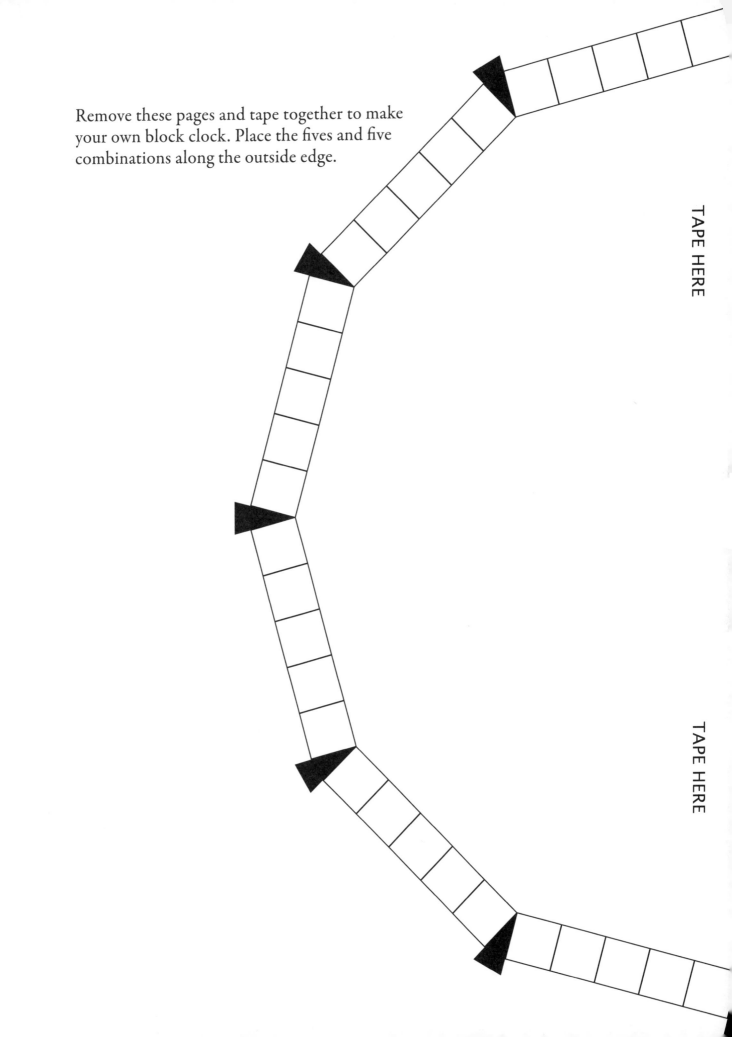

Remove these pages and tape together to make your own block clock. Place the fives and five combinations along the outside edge.

TAPE HERE

TAPE HERE

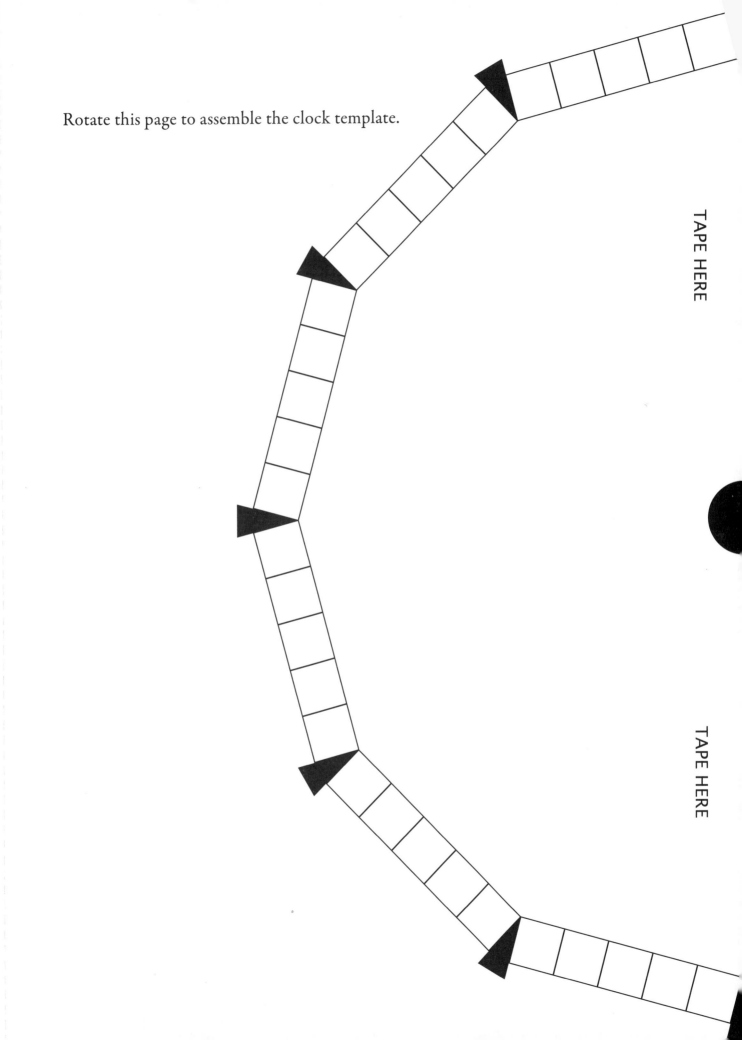

Rotate this page to assemble the clock template.

TAPE HERE

TAPE HERE